Metroburbia,

Metroburbia, USA

PAUL L. KNOX

Rutgers University Press
New Brunswick, New Jersey, and London

Library of Congress Cataloging-in-Publication Data
Knox, Paul L.
Metroburbia, USA / Paul L. Knox.
    p. cm.
  Includes bibliographical references and index.
  ISBN 978–0-8135–4356–7 (hardcover : alk. paper)—ISBN 978–0-8135–4357–4
(pbk. : alk. paper)
  1. Suburbs—United States. 2. Suburban life—United States. I. Title.
  HT352.U6K56 2008
  307.760973—dc22   2007048278

A British Cataloging-in-Publication record for this book is available from the British Library.

Visit our Web site: http://rutgerspress.rutgers.edu

Manufactured in the United States of America

CONTENTS

Figures and Tables

Figures

Tables

PREFACE

This book is the immediate outcome of essays that I wrote for *Opolis* ("Vulgaria: The Re-Enchantment of Suburbia," 2005) and *The American Interest* ("Schlock and Awe," Spring 2007). But in broader terms it is the outcome of a combination of academic interest and personal experience. Having long pursued an interest in urban social geography and in the built environment from a social science perspective, I found myself in the privileged position of Dean of the College of Architecture and Urban Studies at Virginia Tech for almost nine years (1997–2006). During that time my involvement with design professionals and access to planners, builders, developers, and community groups gave me additional insight and somewhat different perspectives on the production and meaning of the built environment. Office visits, site visits, and informal conversations gave depth and texture to issues that had hitherto been, for me, mainly empirical or theoretical in tenor. They also taught me that a summative view is neither possible nor desirable and that a purity of critique is illusory.

This book seeks to bring these perspectives together, weaving data, anecdote, and analysis within an overall framework that is informed by social theory. It offers a reinterpretation of the history of metropolitan form, explores the interdependence of demand- and supply-side factors in the production of the contemporary residential fabric of metropolitan America, and points to the social and cultural significance of the outcome as moral landscapes.

I have been encouraged and assisted by Robert Lang and by other colleagues at Virginia Tech: Katrin Anaker, Kelly Beavers, Elisabeth Chavez, Dawn Dhavale, Karen Danielsen, Jennifer LeFurgy, Asli Ceylan Oner, and Lisa Schweitzer. I would also like to thank and acknowledge the following for permission to use or adapt graphic material: Corbis (Figure 2.1); Robert Lang, Virginia Tech (Figures 3.1, 3.3, 3.10, 3.11, and 8.3); Heike Mayer, Virginia Tech (Figures 3.2 and 3.4); David Theobald, Colorado State University (Figure 3.5); the Brookings Institution (Figures 3.6, 3.7, and 3.8); Robert Lang and Arthur C. Nelson, Virginia Tech (Figure 3.9); Anne-Lise Knox (Figures 4.1, 4.2, and 5.1); Toll Brothers (Figures 4.4 and 8.4); SRI-BC (Figure 7.1); the editors of *Opolis* magazine (Figures 7.2, 7.3, and 7.4); Claritas, a service of the Nielsen Company (Figure 7.5); and The Mall at Millenia (Figure 8.5).

Metroburbia, USA

CHAPTER 1

Introduction

Let's begin with the big picture. American society has been going through some important transformations (along with much of the European Union and several other advanced economies). First, the old economy, based on manufacturing industries, is being displaced by a "new economy" based on digital technologies, biotechnology, and advanced business services. It is an economy that is increasingly dominated by large transnational corporations and intimately tied to complex flows of information and networks of commodity processing, manufacture, and sales that are global in scope. Second, as the economic and occupational structure of the country has changed, so has the distribution of income. America has become increasingly polarized in terms of income and wealth, with more millionaires and more in poverty. Meanwhile—and third—the nation's demographics have also changed: baby boomers are reaching retirement age, GenXers are moving into their mid-thirties, and those of Generation Y—the "Echo Boomers" or "Net Generation" born between 1980 and 1995—are beginning to leave their mark on popular culture and social values.

Fourth, new political sensibilities have emerged, not least in terms of the shift from the egalitarian liberalism that sprang from the New Deal to the neoliberalism unleashed by the Reagan administration. Public funds for almost everything except security have decreased in the wake of grassroots resistance to taxation. States and municipalities, no longer with the capacity to provide or maintain a full range of physical infrastructure or to effectively manage economic development and social well-being, have either abrogated many of their traditional responsibilities or become dependent on public-private partnerships to meet them. Fifth, American society has become more materialistic, more self-oriented, and more narcissistic in its cultural sensibilities. The combination of competitive consumption, a tabloidized infotainment culture, and the dissolution of traditional sociospatial patterns has begun to empty out social relations to the point where some commentators have begun to talk about a "postsocial" world.[1] People's sense of community and sense of place have become so attenuated that "community" and "neighborhood" have become commodified: ready-made accessories furnished by the real estate industry. Confronted with the flattening and speeding-up of the world of popular and material culture, Americans have turned increasingly to an experiential culture of spectacle and nostalgia–better still, spectacle *and* nostalgia in themed settings. People's own identities have at the same time become increasingly equated with consumption and in particular the acquisition of affordable luxuries. Buying things has become both a proof of self-esteem and a means to social acceptance.

All these changes are being transcribed into America's settlement patterns, its cities and metropolitan regions. The traditional dynamics of urban development have been displaced as economic restructuring, digital telecommunications technologies, demographic shifts, and neoliberal impulses have given rise to new urban, suburban, and exurban landscapes. There have been massive changes in real estate investment in the United States in the past quarter century, fueling the emergence of a "New Metropolis" that is characterized by hopscotch sprawl and the proliferation of off-ramp subdivisions; and by polycentric networks of edge cities, urban realms and corridors, exurbs, boomburbs, and micropolitan centers.[2] Urban regions have been extended and restructured to accommodate increasingly complex and fragmented patterns of work and residence, while the political economy of metropolitan America has changed significantly in character in response to socioeconomic realignments and cultural shifts.

These trends severely challenge the ideas about urban dynamics, form, and structure that for so long have been staples in the literature of architecture, planning, and urban geography. As Edward Soja points out in *Postmetropolis*, "In the traditional discourse, the regional morphology of cityspace was seen most broadly as a product of the continuous interplay of centrifugal and centripetal forces emanating from a dominant and generative 'central city.' "[3] Today, metropolitan space has been decentered, with most of the economic dynamism, cultural symbolism, and political ideology emanating from the polycentric periphery. If the industrial metropolis of the nineteenth century was the crucible and principal spatial manifestation of what Ulrich Beck has dubbed the "first modernity," contemporary metropolitan America may be viewed as one of the principal spatial manifestations of a "second modernity," in which the structures and institutions of nineteenth-century modernization are both deconstructed and reconstructed through a more reflexive form of modernization, whereby modernity has begun to modernize its own foundations with a global, rather than a national, frame of reference.[4] As in the "Mega-City Regions" of Europe, the dominant new form of urbanization in America consists of polycentric networks of up to fifty cities and towns, physically separate but functionally networked, clustered around one or more larger central cities, and drawing enormous economic strength from the "new economy."[5] America's metropolitan areas are coalescing into vast, sprawling regions of "metroburbia":[6] fragmented and multinodal mixtures of employment and residential settings, with a fusion of suburban, exurban, and central-city characteristics.

In metroburbia, the leading edges of new development reflect America's contemporary social values and cultural sensibilities. After all, it is in the developers' best interests to make sure they do. This means that scale and spectacle, consumption, and nostalgia (in the form of themed simulations) are all prominent attributes. Take, for example, one of the newest forms of real estate investment: lifestyle centers like Kierland Commons and Desert Ridge Marketplace, on the edge of Phoenix; Santana Row, San Jose; and The Shops at Briar Gate, Colorado Springs. A "boutique" retail concept driven by the booming luxury goods market, lifestyle centers tend to be smaller than traditional malls and do not have anchor stores.[7] Rather, they mimic old-fashioned Main Street settings, with tree-lined sidewalks and lampposts, manicured shrubbery, made-up street names, and plenty of free parking. Besides containing retail chains like Banana Republic, Burberry, Chico's, Coach, Diesel, Gucci, Sharper Image, and Talbots, they also offer movie theaters, upscale restaurants, fitness clubs, outdoor cafés, and street entertainment. According to the International Council of Shopping Centers,[8] there were just 30 lifestyle centers in the United States in 2002, but by 2007 more than 160.

San Jose's Santana Row contains ninety-five specialty shops and restaurants, with a farmers' market on Sundays. The development stretches 1,500 feet—equivalent to three or four city blocks—an ideal distance for a nice stroll but not too long a walk. Above the stores are the wrought-iron balconies of the housing units on the third floor. With plazas, fountains, and street furniture, and outdoor heat lamps to ward off the chill for people who want to linger late into the evening, lifestyle centers are conducive to casual browsing and people watching. But they are privately owned spaces, carefully insulated from the uncertainties of public life. Desert Ridge, for example, has a rigorous code of conduct, posted beneath its store directory. The list of forbidden activities includes "non-commercial expressive activity," "excessive staring," and "taking photos, video or audio recording of any store, product, employee, customer or officer."[9]

Residential development reflects the same approach, with private, master-planned developments having become developers' preferred product line. Planned developments currently contain about 12 percent of all housing units in the United States but constitute roughly 30 percent of new housing starts and 40 percent of new single-family home sales.[10] Maple Lawn, in Howard County, Maryland, midway between Baltimore and Washington, is a typical example. The 600-acre site on former pastureland consists of six different neighborhoods, a "downtown" shopping district with room for a million square feet of office space, and a community center with conference rooms, an events room, an activity room for children, a fitness center, an Olympic-sized pool, tennis courts, and a multipurpose indoor court. The six neighborhoods include a mixture of housing of different types and price ranges, including condominiums, townhomes, "manor homes," "village homes," and "estate homes," all built in neotraditional style to evoke the atmosphere of a small town. The developer promises a "walkable lifestyle" with "top-rated schools" within walking distance of the development, and has dedicated a site in the downtown area for a day care center and preschool. A homeowners' association supervises the day-to-day management of Maple Lawn's residential facilities and is responsible for the maintenance, safety, and upkeep of the community center and all of the development's parks and greens. The association is also responsible for organizing and sponsoring community events, addressing residents' concerns, and enforcing the developer's covenants and design criteria. A second organization, the Maple Lawn Commercial Owners Association, supervises the day-to-day management of the development's commercial facilities.[11] When Maple Lawn was under construction during the peak of the mid-2000s housing boom, demand was so great that homes were sold by lottery.

As typical examples of developments at the leading edges of metroburbia, these developments reflect changing ideals in urban design and the changing aspirations of consumers as well as changing imperatives in real estate development. Laden with layers of symbolic meaning, these everyday landscapes— including the people who inhabit them, their comportment, their clothes, their "stuff"—also amount to moral geographies that both echo and tend to reproduce society's core values. In America, nothing is more central to these values than the American Dream of upward social mobility through ingenuity and hard work, rewarded by steadily improving prosperity and a share in the property-owning democracy. This book is about the way that successive iterations of the Dream have been translated into suburban (and, now, metroburban) landscapes. The central argument is that these landscapes are a product of the interplay of the development industry, design professionals, and affluent consumers with distinctive aspirations and dispositions.

The built environment is, to borrow Umberto Eco's term, an "open text"— an endlessly interpretable series of landscapes.[12] American suburbia has of course already been endlessly interpreted, but, following the canon of architecture, planning, and urban studies, overwhelmingly in terms of idealistic assumptions and deterministic concepts. Thus, a great deal of conventional wisdom has it that American suburbanization is best understood in terms of the ideals of progressive arcadian utopias that are rooted in the Jeffersonian Arcadian Myth or the Frontier Myth.[13] Equally, a great deal of conventional wisdom is framed in rather deterministic terms, focusing on the existence of abundant, cheap land in combination with cheap transportation and lax regulation. In many accounts, therefore, contemporary suburbia is a creature of deep-rooted arcadian and/or pioneer values coupled with (subsidized) automobile dependency and a permissive approach to land use and real estate development.

In this book, suburbanization is reinterpreted as an expression of modernity, focusing on suburbia's roles in terms of consumption and, in particular, on the enchantment that is necessary to sustained consumption and capital accumulation under successive phases of political-economic development. This is not to deny the influence of progressive, arcadian ideals on the discourse of urban design. Nor is it to deny the past influence of federal subsidies or the contemporary influence of cheap gasoline and the shortcomings of land use planning. Rather, the intention is to address the "open text" of contemporary American suburbia in terms of a complementary view that emphasizes the interdependence of consumption and production within the political economy of modern urbanization. This requires an approach that recognizes the nature of the built environment as simultaneously dependent and conditioning, the outcome and mechanism of the dynamics of investment, production, and consumption.

Economic and social relations are constituted, constrained, and mediated through space. At the same time, the built environment is continuously restructured in response to processes of production and consumption and their attendant economic and social relations. The modern metropolis, initially a product of the political economy of the manufacturing era, has been thoroughly remade in the image of consumer society. Producers, for their part, have developed new product lines in response to changing technologies, building systems, regulatory environments, financial systems, and consumer demand. Consumers, meanwhile, have developed new preferences and priorities in response to dramatically changing physical environments, social structures, and patterns of disposable income. Design professionals and exchange professionals have had to adapt to a neoliberal political economy in which progressive notions of the public interest and civil society have been all but set aside. In the process, suburbs have been reconceived and re-formed: from intellectual utopias to bourgeois utopias to democratic utopias to degenerative utopias to conservative utopias, each with a distinctive physical form and moral landscape.

The landscapes at the leading edges of the New Metropolis are important because they accommodate the newest—and for the most part the most affluent—elements in the social ecology of metropolitan America. They are important because they establish the physical framework for the evolution of future urban social ecologies. They are "the backdrop for the showtime of programmed distraction, a system of commerce, a sound stage for affluent lifestyle, a pragmatic infrastructure that sustains flights of individual desire."[14] They are also important as crucibles of consumer culture and as places for transforming and reaffirming class identities. As mute manifestations of ideology and power, they are capable of multiple readings; but developers, realtors, builders, architects, urban designers, and homeowners' associations labor mightily to ensure that some meanings are closed off and that others are promoted through advertising copy, professional and trade magazines, and neighborhood covenants, controls, and restrictions. Landscapes are communicative of identities and community values; they symbolize and insinuate political and moral values as well as create and convey social distinction. The aesthetics of suburban landscapes act as a subtle but highly effective mechanism for exclusion.

Space and Society in Metroburbia

The relationships among social change, consumption, identity, class, space, landscape, and power have been the subject of an extensive literature and a rich array of concepts and theories in social science. To cite just a few influential authors: Jean Baudrillard emphasized the importance of the symbolic component of consumption and its role in class differentiation and intra-class social rivalry. His analysis of codes of consumption has pointed to the importance of simulation—in themed shopping districts, festival settings, and simulated communities, for example—in contemporary culture. Guy Debord elaborated the overarching concept of the spectacle to describe consumer society and the ubiquitous packaging, promotion, display, and media representation of commodities. Others have pointed to the ways in which power and authority become stabilized and legitimized through codes of consumption and the symbolic content of landscape. Michel de Certeau emphasized the "secret murmurings" of cues and codes embedded in the everyday experience of place and space: the underpinnings of inclusion and exclusion and the basis for people's negotiation of identity. Pierre Bourdieu explored the ways in which people's everyday experience results in a distinctive "habitus"—a set of structured predispositions and ways of seeing the world. Michel Foucault pointed to the way that power and authority operates through "normalizing regimes" of social and spatial practices; and Roland Barthes explored the ways in which innocent-seeming social symbols form "codes of domination" that sustain authority—sets of meta-signifiers he called "mythologies."[15]

A common theme in this literature is that our experiences of material and social worlds are always mediated by power relationships and culture. "Social" issues of distinction and "cultural" issues of aesthetics, taste and style cannot be separated from "political" issues of power and inequality or from "gender" issues of dominance and oppression. Another common theme is that urban spaces are created by people, and they draw their character from the people that inhabit them. As social groups occupy urban spaces, they gradually impose themselves on their environment, modifying and adjusting it as best they can to suit their needs and express their values. Yet at the same time people themselves gradually accommodate both to their physical environment and to the people around them. This sociospatial dialectic is a central theme of urban social geography.[16] Neighborhoods and communities are created, maintained, and modified; the values, attitudes, and behavior of their inhabitants, meanwhile, cannot help but be influenced by their surroundings and by the values, attitudes, and behavior of the people around them. At the same time, the ongoing processes of urbanization make for a context of change in which economic, demographic, social, and cultural forces are continuously interacting with these urban spaces. There is, then, a continuous two-way process in which people create and modify urban spaces while at the same time being conditioned in various ways by the spaces in which they live and work. Similarly, cities are constantly made and remade in response to all manner of economic, social, cultural and political forces; but at the same time they mediate, constrain, and inflect particular processes in distinctive ways.

These ideas are central to structuration theory, which addresses the way in which everyday social practices come to be structured across space and time. Developed by Anthony Giddens,[17] structuration theory holds that human landscapes "are created by knowledgeable actors (or agents) operating within a specific social context (or structure). The structure-agency relationship is mediated by a series of institutional arrangements that both enable and constrain action. Hence three 'levels of analysis' can be identified: structures, institutions, and agents. Structures include the long-term, deep-seated social practices that govern daily life, such as law and the family. Institutions represent the phenomenal forms of structures, including, for example, the state apparatus. And agents are those influential human actors who determine the precise, observable outcomes of any social interaction."[18]

We are all actors, then (whether ordinary citizens or powerful business leaders, members of interest groups, bureaucrats, or elected officials), and all part of a dualism in which structures (the communicative structures of signification and symbolism as well as formal and informal economic, political, and legal structures) frame and enable our behavior while our behavior itself reconstitutes, and sometimes changes, these structures. Furthermore, structuration theory recognizes that we are all members of various networks of social actors: organizations, interest groups, neighbors, social classes, and so on. The way we generate meaning within these systems is rooted in routinized day-to-day practices that occupy a place in our minds somewhere between the conscious and the unconscious. Recursivity, the continual reproduction of individual and social practices through routine actions, contributes to social integration, the development of social systems and structures among agents in particular locales.

From Weberian sociology we are reminded that some actors and institutions are often more significant than others. Weberian analysis seeks to explain how the outcomes of competition between conflicting and competing social groups are influenced by institutional organization and key mediating professionals. The development of this approach in relation to contemporary processes of urbanization can be traced to the work of Ray Pahl, who argued for a focus on mediating professionals such as planners, mortgage managers, and realtors.[19] Each of these sets of key actors develops a distinctive professional ideology and value system as a result of recruitment, education, and professional reward systems. Drawing on these ideologies in their day-to-day decision making and their interpretations of social needs, priorities, and market forces, these key actors can exert a decisive influence on urban outcomes, sometimes subtly, sometimes not. In some cases, their influence can be indirect, influencing people's sense of possibilities without ever becoming directly involved in an issue.

The Enchantment of Material Things
Another strand of neo-Weberian theory that is useful in understanding the way that successive iterations of the American Dream have been translated into suburban landscapes comes from Colin Campbell's reinterpretation of the relationship between capitalism and the Protestant Ethic.[20] Max Weber's view of nineteenth-century capitalism was of a rationalized system infused with a Calvinist spirit, concerned with economic success but coldly ascetic, antithetical to mystery, romance, and enchantment. Looking back from the late twentieth

century, Campbell charted the transition from this practical rationality to a "spirit of modern consumerism" that had its origins in people's need to establish social metrics of good taste. Refinement and good character, Campbell argues, was initially attributed to people who sought beauty and goodness and derived pleasure from them. Soon, pleasure-seeking came to be tied to the consumption of beautiful luxury goods. People's lives became infused with illusions, day-dreams, and fantasies about consumer objects. Thus emerged the spirit of modern consumerism, characterized by Campbell as a "self-illusory hedonism." Under the spell of self-illusory hedonism, people constantly seek pleasure, enchanted by a succession of objects and ideas, always believing that the next one would be more gratifying than the previous one. This is the basis of a "romantic capital-ism," driven more by dreams and fantasies than a Protestant work ethic.

Romantic capitalism blossomed in the 1950s with a postwar economic boom that was boosted by the widespread availability of credit cards. Traditional identity groups based on class, ethnicity, and age began to blur as people found themselves increasingly free to construct their identities and lifestyles through their patterns of consumption. In addition to the traditional business of posi-tional consumption, members of new class fractions and affective "neotribal" groupings sought to establish their distinctiveness through individualized pat-terns of consumption.[21] Consumption eclipsed production as the most important arena for social, cultural, and political conflict and competition. Thanks to Fordism,[22] consumers' dreams could be fulfilled more quickly and more easily. Enchantment sprang from the affordability and choice resulting from rationaliza-tion and mass production. But this led inevitably and dialectically to disenchant-ment as novelty, exclusivity, distinction, and the romantic appeal of goods were undermined by mass consumption. To counter this tendency, product design and niche marketing, along with the "poetics" of branding, have become central to the enchantment and reenchantment of things.[23] As George Ritzer has pointed out, enchantment is also ensured through a variety of specialized settings—"cathedrals of consumption"—geared to the propagation and facilitation of consumption: shopping malls, chain stores, catalogs, franchises and fast food restaurants, Internet and TV shopping, cruise ships, casinos, and themed restaurants.[24] Ritzer, following Baudrillard, Debord, and others, points to the importance of spectacle, extravaganzas, simulation, theming, and sheer size in contemporary material cul-ture, and argues that they are all key to enchantment and reenchantment in the consumer world.

The Structures of Building Provision

With these ideas on the complexity of sociospatial dialectics in mind, metroburbia can be seen as both a structured and structuring environment, and the production and consumption of the built environment can be seen not simply as a matter of supply and demand but also as a function of time- and place-specific social relations that involve a variety of key actors and institutions, includ-ing landowners, investors, financiers, developers, builders, design professionals, business and community leaders, government agencies, homeowner associations, and consumers. At the same time, it is clear that these relations among key agents and institutions need to be understood in terms of their linkages with the broader sweep of economic, social, cultural, and political change. These sets of relations represent "structures of building provision"[25] through which we can better understand the evolution and character of metroburbia.

The suburban residential development process has been described as "a sort of three-dimensional spider web that can be moved by impact in any corner."[26] Every project involves a variety of agents, each with their own objectives, motivations, resources, and constraints, and many of them connected with one another in several different ways. In every metropolitan region there are thousands of major landowners, and hundreds of developers and builders. Some agents act for themselves within the web of the development process; others represent groups of people, large corporations, or public agencies. Some agents may play more than one role at a time. Landowners may be actively involved in subdividing and building, for example, while city governments may act as both regulators and entrepreneurs.

This complexity notwithstanding, and recognizing that each metropolitan region has its own political culture and market conditions, it is nevertheless useful to enter a few preliminary and general observations about some of the key elements in the structures of building provision, especially with regard to their motives and constraints.[27] Landowners stand at the beginning of the chain of events involved in urban development, and they can influence patterns of growth at the leading edges of metropolitan regions through the size and spatial pattern of parcels of land that are sold to developers and through the conditions that they may impose on the nature of subsequent development. In terms of the size and spatial pattern of land parcels, much of course depends on the overall pattern of land holdings. The large *ranchos* and mission lands around Los Angeles, for example, have formed the basis of extensive tracts of uniform suburban development, while in eastern cities, where the early pattern of land holdings was fragmented, development has been more piecemeal. Because of the structure of the tax system, however, it is often preferable for landowners to sell smaller parcels over an extended period. This consideration applies especially to commercial farmers, a group that is crucial to the land conversion process at the urban fringe. Speculators are another important category of landowners—especially the bigger players who rely not merely on an ability to anticipate trends but who also hope to influence or engineer change for their own benefit. They may attempt, for example, to influence the route of a highway or the location of a rapid transit stop, to change a land-use zoning map or master plan, or to encourage public expenditure on particular amenities or services. This points to an important dimension of structures of building provision throughout metroburbia: "growth-machine" coalitions of local real estate, finance, and construction interests that seek to propagate an ideology of growth and consumption as well as engaging in tactical politics around local government land-use regulation, policy, and decision making.

Developers and builders, of course, are central to growth machine politics. Because it is developers who typically must decide upon the type of project to be undertaken on a particular site, they can fairly claim to be the single most important group in shaping metroburbia, inscribing their judgment and interpretation onto the landscape. The fundamental role of developers is in deciding upon the nature and form of new projects, platting large parcels of land into smaller lots, installing the infrastructure of streets, sewer and water mains, gas and electric lines, and selling the lots to builders. Many development companies, however, have extended their activities well beyond this business of subdivision to include land assembly and speculation, design, construction, and marketing. For most residential development, the overarching criterion for developers is that of site costs. But different-sized firms have different constraints and opportunities. Suburban development in the 1950s, 1960s, and 1970s was dominated everywhere by small- and medium-size companies that tended to opt for what was easiest to produce and what was the safest bet in terms of effective demand—two- and three-bedroom single-family suburban housing for middle-middle class households. But in the 1980s, when developers and their marketing consultants caught up with social shifts that had made the "typical" middle-class household a sociodemographic minority, they began to cater to affluent singles, divorcees, retirees, and the growing number of upper-middle income households, adding luxury condominiums, townhouses, artists' lofts, and private master-planned communities to their product lines.

Vertical and horizontal integration in the industry meanwhile resulted in ever-larger companies that have come to account for ever-greater shares of new home construction. These larger companies have a compelling need to acquire land continually in order to ensure a supply of developable land and keep their organization fully employed. They also need land in parcels large enough for them to exploit economies of scale. As a result, most of them have extensive land banks, often searching for suitable land before it has been put on the market (and before any thought has been given to project conceptualization): a strategy known in the trade as bird-dogging.

For these large companies, the most profitable markets are the middle- and upper-middle classes. The boom in the "new economy," driven by the growth of industries based on digital technologies, biotechnology, and advanced business services, fueled a housing boom, especially at the top end of the market. Then, with the collapse of the dot-com speculative bubble in 2000–2001, property markets received a further boost as the built environment became a refuge for capital. Big capital needed a safe refuge, while the sobering collapse of the heady dot-com boom made real estate an attractive investment for affluent households. People traded up as fast as they could, aided by a credit industry that became increasingly competitive and increasingly lenient, offering all sorts of packages to make big mortgage repayments easier for buyers to contemplate.

What affluent households wanted for their money was a home worthy of their repayment burden. In the sprawling, fragmented, and polycentric New Metropolis, with cloned off-ramp developments and themed master-planned communities with made-up place names, the old real estate aphorism—location,

location, location—became less relevant for many households, especially those able to telecommute for part of the week. An address used to count for much more when cities were more compact and coherent: people had a clear mental map, with clear social gradations. Now, people's mental maps are fuzzy and piecemeal. The new mantra is size, size, location; or just size, size, size. Conspicuous construction has become an important precondition for conspicuous consumption, and suburban social space has become irradiated by bigness and bling.[28]

The Enchantment of Suburbia

The built environment of suburbia, like many other components of material culture in romantic capitalism, has been the subject of recursive and overlapping sequences of enchantment, disenchantment, and reenchantment. While the dreams, increasing affluence, and reflexive abilities of consumers have driven the demand side of the serial enchantment of suburbia and the increasing size and scope of developers has driven the supply side, demand and supply are mediated by a whole spectrum of "exchange professionals," many of whom are complicit in the processes of enchantment and reenchantment. These key actors include mortgage financiers, realtors, surveyors, market analysts, advertising agencies, lawyers, appraisers, property managers, engineers, and ecologists. Most important in terms of the enchantment and reenchantment of the built environment, however, are design professionals: architects, planners, urban designers, and landscape architects. As key arbiters of style in modern society, design professionals are in a powerful position to stimulate consumption merely by generating and/or endorsing changes in the nuances of building and landscape design. Meanwhile, by virtue of the prestige and mystique socially accorded to creativity, design professionals add exchange value to the built environment through their decisions about design.

More broadly, we look to design professionals to draw on their particular skills to translate social and cultural values into material form: "Their products, their social roles as cultural producers, and the organization of consumption in which they intervene create shifting landscapes in the most material sense. As both objects of desire and structural forms, their work bridges space and time. It also directly mediates economic power by both conforming to and structuring norms of market-driven investment, production, and consumption."[29] Design professionals are cultural transistors: amplifiers and rectifiers of intellectual, social, and cultural signals; at once both products and carriers of the flux of ideas and power relationships inherent to particular stages of metropolitan development. Major shifts in architectural style can be seen as a dialectical response to the evolving *zeitgeist* of urban-industrial society. In this regard, design professionals are particularly influenced by broader artistic, literary, and intellectual movements. In terms of suburban domestic architecture, for example, the arcadian and utopian ideals of the nineteenth-century American Renaissance have been especially important, echoing all the way down to many of the private master-planned communities of contemporary metroburbia.

While they must work within certain parameters set by clients, politicians, legal codes, and so on, design professionals inevitably bring their distinctive professional ideologies to bear. The professional ideology and career structure that rewards innovation and the ability to feel the pulse of fashion also serves to promote enchantment and, therefore, the circulation of capital. Without a steady supply of new fashions in domestic architecture (reinforced by innovations in kitchen technology, heating systems, and so on), the filtering mechanisms on which the whole owner-occupier housing market is based would slow down to a level unacceptable not only to builders and developers but also to other exchange professionals and the whole range of financial institutions involved in the housing market. The upper-middle classes, in short, must be encouraged to move from their comfortable homes to new dwellings with even more "design," "convenience," and "luxury" features in order to help maintain a sufficient turnover in the housing market. Hence the desperate search for successful design themes to be revived and "re-released," just like the contrived revivals of haute couture and pop music. In metroburbia, the process has advanced to the stage where many themed developments resemble small chunks of Disneyland: style for style's sake, the *zeit* for sore eyes. In many cities, new housing for upper-income groups is promoted through annual exhibitions of "this year's" designs, much like the Fordist automobile industry's carefully planned obsolescence in design.

The Plan of the Book

This book is broadly organized around the concept of the structures of building provision, looking successively at the roles of key actors and institutions in the production of the cultural landscapes that dominate the leading edges of the New Metropolis—the landscapes of what I call "Vulgaria." Vulgaria is not always an extensive tract or a coherent zone or sector of metroburbia but, rather, a suffusion of distinctive and influential landscapes centered around the themed and packaged upper-middle class residential developments and commercial megastructures of suburbia and exurbia. Chapter 2 puts current trends in suburban development in historical context, emphasizing the role of intellectuals, designers, developers, builders, and realtors in the serial enchantment of successive phases of suburban development as new "spaces of hope" infused with utopian characteristics. Chapter 3 locates contemporary upscale suburbia within the changing social and spatial structures of metropolitan America: the "New Metropolis." Chapters 4 through 7 deal with the key actors and institutions in the development of these reenchanted suburbs: builders and developers (chapter 4), design professionals and consultants (chapter 5), institutions of governance (chapter 6), and consumers (chapter 7). The concluding chapter characterizes the conservative utopia of Vulgaria and discusses its landscapes as moral geographies. Vulgaria has naturalized an ideology of competitive consumption, moral minimalism, and disengagement from notions of social justice and civil society— the peculiar mix of political conservatism and social libertarianism that is the hallmark of contemporary America.

CHAPTER 2

Prelude

THE SERIAL ENCHANTMENT OF SUBURBIA

The settings that provide the arenas for the contemporary social geography and material culture of metroburban America are the legacy of serial enchantment and disenchantment. Each phase in this history has been shaped to some degree by intellectuals, designers, developers, builders, and realtors, as well as by the aspirations of households and the resources, economic climate, and technologies of the time. The phases have overlapped, and the influence of some of the key actors echoes through several phases, so that there is no neat way of classifying them.[1] A persistent theme in this history, however, has been the notion of arcadian suburban settings of one form or another as an ideal, a manifest destiny of American society. More specifically, the serial enchantment of successive phases of suburban development has rested on the prospect of new "spaces of hope"[2] infused with utopian characteristics. On the other hand, the eventual disenchantment with particular phases of suburban development has typically invoked a dystopian discourse: an essential ingredient in the dialectical materialism of metropolitan America.

The cumulative legacy of this serial enchantment and disenchantment is central to the condition of contemporary metroburban America: it is ingrained in its political economy, its material culture, and its social geography. The conservative commentator David Brooks has characterized this legacy, as reflected in both the intellectual and popular literature of the post-World War II era, as "a comfortable but somewhat vacuous realm of unreality: consumerist, wasteful, complacent, materialistic, and self-absorbed. Sprawling, shopping, Disneyfied Americans have cut themselves off from the sources of enchantment, the things that really matter. They have become too concerned with small and vulgar pleasures, pointless one-upmanship, and easy values. They have become at once too permissive and too narrow, too self-indulgent and too timid. Their lives are distracted by a buzz of trivial images, by relentless hurry instead of genuine contemplation, information rather than wisdom, and a profusion of superficial choices."[3] Disenchantment, then. Sprawl, placelessness, punishing commutes, loss of community, fear, and a mean-spirited, neoliberal political economy are the hallmarks of the cumulative legacy of suburban development. How did we reach this state?

Serial Enchantment

The initial charm of suburban living derived not so much from the attractions of the suburbs themselves as from their relative appeal in comparison with the uncontrolled turmoil of the industrial cities of the mid-nineteenth century. The wealthy had already been sending their families away to the countryside for the uncomfortable and epidemic-prone summer months, but the sudden proliferation of all kinds of noxious industrial land uses, together with the arrival of thousands of immigrants and their subsequent emergence as a massive and threatening proletariat, soon gave upper- and upper-middle income families—including the rapidly growing new class of affluent white-collar workers—a strong motivation for moving permanently as far as possible away from the central city. A few families, wealthy enough to be able to afford private carriages or hire hackney carriages, had moved out to exclusive exurban settings at the first signs of industrial squalor. But it was not possible for others to join them until the development of horse-drawn omnibus systems, horsecars (light rail systems drawn by horses), and short-haul passenger railroad routes.[4] When this happened, the character of the suburbanization that took place was deeply influenced by intellectuals,' architects,' and planners' (the categories are not always mutually exclusive) visions of a suburban utopia.

Intellectuals' Utopias

American intellectuals and design professionals were especially influenced by the literary culture of the so-called American Renaissance. Dating from the 1830s to roughly the end of the Civil War, the American Renaissance was rooted in the works of Henry Wadsworth Longfellow, Oliver Wendell Holmes, and James Russell Lowell, who sought to create a genteel American literature based on foreign models. One of the most important influences in the period was that of the Transcendentalists, centered in the village of Concord, Massachusetts, and including Ralph Waldo Emerson and Henry David Thoreau. The Transcendentalists were explicitly anti-urban, viewing cities as diseased, dangerous, and even infernal. Emerson had drawn, in the 1830s, on the European Romantics' notion of the pastoral ideal in arguing for settlements that incorporated the benefits of both city and country. Thoreau, a disciple of Emerson, famously observed that in industrial cities "the mass of men lead lives of quiet desperation." He advocated accessibility to Nature as a spiritual wellspring for city dwellers in his book *Walden* (1854).[5] By the mid-nineteenth century, Americans had come to think of their relationship with Nature and the Great Outdoors as something distinctively "American." The historian Frederick Jackson Turner advanced the influential idea that the "frontier experience" was the single most significant factor in determining the American character, providing a release valve that allowed upward mobility due to the development and improvement of virgin land,[6] and it became broadly understood that "access to undefiled, bountiful and sublime Nature is what accounts for the virtue and special good fortune of Americans."[7] Meanwhile, a central aspect of nineteenth-century anti-urban ideology rested on a fear of the loss of patriarchal control over women amid the economic and social turbulence of industrializing cities. Writers at the time rarely acknowledged this fear in any explicit way, preferring the implicit promise of patriarchal control that came with their idealized vision of the private domesticity of arcadian settings.

Against this backdrop, the American Renaissance propagated the ideal of settings in which man and Nature could achieve a state of balance–what landscape architect Leo Marx has described as a "middle landscape" of pastoral and picturesque settings.[8] At the same time, the intellectuals of the period emphasized the moral superiority of domesticity and the virtues of republicanism and sanitary reform.[9] This led to the widespread acceptance of a vision of ideal suburban settings that combined the morality attributed to Nature with the enriching and refining influences of cultural, political, and social institutions. Progressive intellectuals like Andrew Jackson Downing advocated a program of "popular refinement" involving the creation of a whole series of institutions and settings such as public libraries, galleries, museums, and parks, in order to bring out the best in "ordinary" people.[10]

In practice, the discourse surrounding the actual development of suburbia was complex and fluid. In their study of Bedford, in Westchester County, New York, James and Nancy Duncan showed how nineteenth-century exurbanites brought with them an ambivalent anti-urbanism and romantic images of the countryside as a site of stable and healthy social relations, while the subsequent development of suburban Bedford drew on a shifting complex of narratives and ideologies: "The discourses include Arcadian ecology; the romantic idea of wilderness; Thoreauean transcendentalism; a Jeffersonian agrarian vision of the virtuous yeoman farmer; aristocratic stewardship of land; rural republicanism; the English pastoral ideal including the country cottage, wild garden, and country house; Puritan New England democratic values; and historic preservationism."[11] Discourses such as these have evolved and combined to create in suburbia a potent moral geography that has resonated through metropolitan development for the past 150 years.

The architects, landscape architects, and planners who translated into built form the ideals implicit in these complex discourses "set about their tasks by combining an intense imaginary of some alternative world (both physical and social) with a practical concern for engineering and reengineering urban and regional spaces according to radically new designs."[12] The histories of these pioneer design professionals and their projects have been written up in detail;[13] the concern here is simply to point to the influence of the seers and pioneers whose legacy has shaped successive phases of enchantment, disenchantment, and reenchantment.

The early seers included Alexander Davis, who designed Llewellyn Park in West Orange, New Jersey, a railway suburb begun in 1853 by Llewellyn Haskell, a successful Manhattan merchant. Llewellyn Park was designed as an exclusive residential enclave and Davis and Haskell together pioneered several key innovations that have become commonplace: encompassing landscaping that was carefully manipulated to give a natural appearance; detailed deed restrictions covering architecture and domestic landscaping; and a homeowners' association to enforce these restrictions and to administer the commonly owned landscaped areas.

Better-known among the early seers—and in fact the single most important figure in American planning, according to urban historian Robert Fishman—was Frederick Law Olmsted. Olmsted saw his work as serving the psychological and social needs of city residents to have access to a naturalistic landscape, a secluded escape from the dirt and noise of the city, a place for leisure and recreation, and an environment that would foster restraint and decorum.[14] New York's Central Park was Olmsted's first opportunity to achieve those goals on any scale. In 1858 he and Calvert Vaux won the competition for the design of the park, which was completed in 1862. Their design included a succession of specific areas for sport, recreation, and culture, all embedded in a picturesque landscape. The park was integrated into the city by means of four avenues laid out with an elaborate system of independent traffic lanes, bridges, and underpasses that were designed not to interrupt the continuity of the landscape. The result was widely acclaimed, and Olmsted went on to design park projects in other cities (including Boston, Brooklyn, Buffalo, Chicago, Detroit, Louisville, Milwaukee, Newark, Philadelphia, and San Francisco) and campuses for the University of California at Berkeley and Columbia University in New York.

Restorative Utopias: Garden Suburbs and Garden Cities. The ideas developed by Olmsted and Vaux in these park projects carried over into their work on garden suburbs. In the tradition of the American Renaissance, they saw planned suburban development as a chance to provide the benefits of city life without the congestion, tumult, noise, crime, and vice, and the benefits of country life without the inconvenience, isolation, and lack of amenities. Their first project, in 1869, was Riverside, a Romantic-styled railway suburb nine miles from downtown Chicago. Like Llewellyn Park, Riverside was lushly landscaped in a park-like setting, allowing Nature to bestow her virtues; its homes were generously proportioned, securing privacy for womenfolk and status for paterfamilias; and its communal open spaces were to be the arenas for the emergence of a progressive, reformed community life. The enchantment of Riverside was underpinned by the existence of a ready-made commuter system—the railroad—and soon other garden suburbs, including Highland Park, Lake Forest, and Winnetka, sprang up around Chicago's local railway network. Over the course of the late nineteenth and early twentieth centuries, similar "picturesque enclaves" emerged around other large cities, catering to upper-middle class railway commuters.[15]

Meanwhile, the broader reform movements of the late nineteenth century prompted others to explore a more ambitious idea altogether: the concept of garden cities that would accommodate not only the upper-middle classes but to the full spectrum of society, with jobs and civic amenities as well as homes: "restorative utopias."[16] In England, Ebenezer Howard distilled and codified the garden city idea in his famous book *To-morrow: A Peaceful Path to Real Reform*, published in 1898 and reissued in 1902 as *Garden Cities of Tomorrow*. Howard's rationale and plans for a garden city drew heavily on philanthropic ideas of the time, together with the socialist anarchism of Peter Kropotkin, the communitarianism of Charles Fourier, the socialist ideals of William Morris, and the aesthetic principles of John Ruskin.[17] Echoing Olmsted and Vaux's rationale for garden suburbs, he portrayed the garden city ideal as incorporating the best of city life (jobs, higher wages, civic amenities, social interaction, etc.) with the best of life in the countryside (clean air, natural beauty, open space)—while avoiding the downside of both (the squalor and pollution of cities; the limited opportunities and poor infrastructure of rural areas).

Howard himself invested in the first true garden city, Letchworth, thirty-four miles from London, founded in 1904. A financial success—for it was built for profit—it was followed by other projects: Hampstead Garden Suburb (in greater London), Welwyn Garden City (near London), and Wythenshawe (near Manchester). The physical designs for these new moral landscapes were inspired by Raymond Unwin and Barry Parker, cousins who grew up near the industrial city of Sheffield. Neither was formally trained as an architect, but both saw their mission as promoting beauty and amenity. Like Howard, they were heavily influenced by William Morris. "They believed that creativity came from an imaginative understanding of the past; that the Middle Ages provided an historic standard; that old buildings grew out of the ground they stood on; that the village was an organic embodiment of the small, personally related community; that the architect and planner were guardians of social and aesthetic life, maintaining and enhancing the traditional values of the community for future generations."[18]

These convictions, together with the practical successes of the English garden cities, resonated resoundingly with intellectuals and developers in the United States. The Russell Sage Foundation developed the first garden city in the United States, Forest Hills Gardens, in Queens, nine miles from Manhattan, in 1911. Designed by Grosvenor Atterbury and Olmsted's son, Frederick Law Jr., Forest Hills Gardens contained a kitsch-like mix of housing laid out with a distinctive neighborhood structure that convinced one of its residents, Clarence Perry, that the layout of a project could, if handled correctly, foster "neighborhood spirit." Perry developed the concept of the neighborhood unit, defined by the catchment area of an elementary school, focused on the school itself, local stores, and a central community space, and bounded by arterial streets wide enough to handle through traffic. As Peter Hall points out, Perry saw the neighborhood unit as an opportunity for social engineering that would assist in nation-building and the assimilation of immigrants, but the idea of neighborhood spirit also went down well with communitarians, and the idea of handling traffic went down well with planners who were beginning to grapple with the implications of the spread of automobile ownership.[19]

During the 1920s and 1930s Perry's neighborhood unit concept was adopted and developed by others—notably Clarence Stein and Henry Wright, in landmark developments that included Sunnyside Gardens (New York), Radburn (New Jersey), Chatham Village (Pittsburgh), and Baldwin Hills Village (Los Angeles). John Nolen, drawing on the ideals of the American Renaissance, sought to revive the association of physical design with the civic ideals in his book *New Towns for Old*.[20] Nolen advocated the pursuit of what he called the "Greek Ideal"—rationality, the study of nature, and celebration of life—through urban design. Meanwhile, garden city ideals were being adopted and sponsored by the Regional Planning Association of America, whose evangelical spokesperson-in-chief was Lewis Mumford. The RPAA's intellectual roots were distinctly radical and owed a lot to Patrick Geddes, an eccentric polymath who taught biology at the University of Dundee in the United Kingdom "and tried to encapsulate the meaning of life on folded scraps of paper."[21] Geddes was particularly inspired by the French sociologist Frederic Le Play and the French geographers Paul Vidal de la Blache and Elisée Reclus, from whom he developed his idea of the "natural" region: cultural landscapes resulting from the harmonious balance of natural resources and human settlement. For Geddes, natural regions offered a framework within which to reconstruct social and political life entirely—an objective whose intellectual antecedents sprang from the anarchism of Michael Bakunin, Pierre-Joseph Proudhon, and Peter Kropotkin.[22] Geddes's own writings tended to be meandering and incoherent,[23] and it fell to Mumford to translate this radicalism to the American context, fusing it with the more restorative gospel of the garden city movement and the conservative social engineering of Perry, Stein, and Wright.[24] Mumford and the RPAA exercised enormous influence on American planning practice and planning ideology from the 1930s onward, not least through the legacy of Franklin Delano Roosevelt's New Deal programs.

Beaux Arts and The City Beautiful.   While all this was unfolding, a much more conservative approach to the built environment had flourished briefly under the leadership of America's architectural avant garde. The Beaux Arts movement took its name from L'Ecole des Beaux Arts in Paris, where, from the mid-nineteenth century, architects were trained to draw on Classical, Renaissance, and Baroque styles, synthesizing them in new buildings that might blend artfully with the significant older buildings that dominated European city centers. Though American cities had no buildings of comparable vintage that needed to be complemented in this way, the Beaux Arts style provided American architects and developers a convenient packaging of High Culture that promised to resolve confusion and uncertainty about how to build in ways appropriate to what Geddes and Mumford called the "neotechnic" era. The idiom was showcased by Daniel Burnham's neoclassical architecture for the World's Columbian Exposition in Chicago in 1893. The temporary structures of the Exposition showed what might be done, and before long Beaux Arts buildings were popping up across America, in suburbs and small towns as well as in city centers.

At the turn of the century the Beaux Arts ideal fused with the reformist ideology of the Progressive Era, giving rise to the City Beautiful movement.[25] It is clear that the City Beautiful movement was an explicit and rather authoritarian attempt to create moral and social order in the face of urbanization processes that seemed to threaten disorder and instability. The thrust of the movement was decisively toward the role of the built environment as an uplifting and civilizing influence. The preferred architectural style was Beaux Arts neoclassical, accompanied by matching statuary, monuments, and triumphal arches—all, if possible, laid out like Burnham's White City at the Chicago Exposition, with uniform building heights and imposing avenues with dramatic perspectives.

In 1901, Burnham collaborated with several others (including Frederick Law Olmsted Jr.) on the McMillan Plan for Washington, D.C. (named after Senator James McMillan, chairman of the Senate Committee on the District of Columbia). The purpose of the McMillan Plan was to rescue the Mall area from the neglected and unfinished framework derived from Pierre Charles L'Enfant's original plan of 1791. The centerpiece of the new plan was the redeveloped Mall and Federal Triangle, with neoclassical buildings along the Mall, a terminal memorial (the Lincoln memorial), a pantheon (the Jefferson Monument), the Memorial Bridge, and a water basin. Although the scheme was not completed until 1922, the plans and sketches provided enough publicity to ensure the immediate future of the City Beautiful movement. Burnham went on to draw up plans for Cleveland (in 1902), San Francisco (1905), and Chicago (1909) before his death in 1912. These plans were also very influential, though little of the San Francisco plan was actually realized. Meanwhile, John Olmsted (Olmsted Sr.'s stepson) devised a plan for Seattle; others worked on schemes for Kansas City, Denver, and Harrisburg, Pennsylvania.

The City Beautiful movement flourished—albeit briefly—because it allowed private enterprise to function more efficiently while symbolizing a noble idealism that was endorsed by the pedigree of Beaux Arts neoclassicism. Over and beyond the good intentions of its leading figures and technicians, "the City Beautiful Movement perfectly fulfilled its true function of matching maximum planning with maximum speculation."[26] Its success did not last long, however, because the dynamics of urbanization were changing: Burnham, Olmsted, and their followers were prescribing neoclassical remedies to industrializing cities. Streetcars and electric power had begun to turn cities inside out, the severe recession of 1893–94 had shaken up relations between classes and introduced profound changes to the social and political organization of the city, and, by the time of Burnham's death in 1912, automobiles were beginning to make their mark.

Realtors' Utopias and the American Dream
Notwithstanding the reforms of the Progressive Era and the interventions of the City Beautiful movement, the increasingly numerous and more affluent

middle-income groups became increasingly frustrated at being cooped up in their cities along with the factories, railroads, warehouses, and "huddled masses" of "ordinary" people. Several decades of development of picturesque enclaves of railway suburbs had firmly established suburbia as a powerful cultural ideal, its enchantment derived from "the capacity of suburban design to express a complex and compelling vision of the modern family freed from the corruption of the city, restored to harmony with nature, endowed with wealth and independence, yet protected by a close-knit, stable community."[27]

There were plenty of plans, plenty of builders and developers ready to execute them, and plenty of land. A quick and inexpensive new building technology—balloon framing[28]—meant that residential construction no longer required a lot of skilled (and expensive) labor. Suburban housing was relatively inexpensive, but there was no means of transportation capable of handling the tens of thousands of potential commuters at affordable prices. Suddenly, in 1888, this tension was relieved by the introduction of a simple but radically effective innovation in urban transportation: the electric streetcar, or trolley.[29]

By 1902, more than 200 American cities had established streetcar systems, with a total 22,000 miles of streetcar tracks. By making it feasible to travel up to ten miles from the downtown district in thirty minutes or so, the streetcar greatly increased the territory available for residential development and opened the way for the growth of streetcar suburbs: the first full flush of what Robert Fishman memorably described as "Bourgeois Utopias." So much land became accessible at once that the price of land was kept down, thus ensuring inexpensive suburban lots. Affordable land, combined with the cheaper operating costs per passenger-mile of the streetcar (because of their large carrying capacity and the efficiency of electric power), ensured that the latent demand of middle-class families to flee the city became *effective* demand.

At last, middle-class families could pursue their dream of bourgeois lifestyles in sanitized arcadias of collective privacy and respectability: "sacred" spaces that contrasted strikingly with the "profane" industrial city. "What had only recently been luxuries for the few became standard services for the many: running water, electricity, gas, paved roads, sewers, police and fire protection, parks and public schools, and a mass transit system that connected the neighborhood to jobs in the factory district and to downtown."[30] Meanwhile, the development of streetcar suburbs provided an outlet for the investment of newly won capital that the new middle classes had at their disposal. "Hence, right from the beginning, the suburban landscape was commodified. . . . Once the ideal had been established, and once the family had been remade to fit the landscape, even as the landscape was remade to fit the family, suburbia exploded, becoming, as it were, the only option for respectable middle-class life."[31]

These commodified bourgeois utopias were also, of course, developers', builders', and realtors' utopias. Real estate, as Thorstein Veblen once wrote, became the "great American game." It was played out within a political economy that was as close to the ideals of classical liberalism as there would ever be: with a broad acceptance of the primacy of individual autonomy, the efficacy of unfettered markets, and the desirability of a non-interventionist, *laissez-faire* state.[32] Developers of streetcar suburbs instantly found a market in the solid middle-income groups, and there was a rush of speculative suburban sprawl as developers and streetcar operators worked together to shape the massive increment of urban growth. Around the fringes of Chicago, for example, some 80,000 new residential lots were recorded between 1890 and 1920 (though many of these lots were not built on for some time afterward). In general, the lots that did get built up were those within walking distance of the streetcar stops, so that the form of American cities in this era was characterized by fingerlike linear extensions from the pre-streetcar core. Within these extensions, development typically took the form of a continuous strip of commercial development and apartment houses lining the streetcar route, with gridded residential streets extending behind for a few blocks. As streetcar systems grew, the streetcar companies sought to encourage still more ridership, and many followed the example set early on by Henry Whitney in Boston, establishing a flat-rate nickel fare. This low fare made streetcar suburbs financially accessible to thousands of lower-middle-class households, leaving the core of the city increasingly to a residual population of low-income households.

In every city there emerged a very close relationship between the construction of mass transportation systems and suburban real estate development. In some cities, transit lines and real estate development were undertaken by the same organization—the streetcar suburb of Shaker Heights in Cleveland is an example.[33] Transit lines often preceded land development, the profits from substantial increases in land value—triggered by imminent access to transit lines—more than offsetting any early losses on the streetcar service. The potential profits from suburban real estate also motivated a great deal of corruption within the nexus of suburban landowners, developers, financiers, and politicians. This, of course, at a time when Progressive Era zealots were focused on reforming central cities.

In most cities, the actual construction of homes was dominated by small businesses. Sam Bass Warner's study of streetcar suburbs in Boston, for example, found that between 1890 and 1900, when the population of suburban Roxbury, West Roxbury, and Dorchester grew from 60,000 to 227,000, some 9,000 builders were active. Most of these builders were operating in their own neighborhood, and 40 percent of them were building within a block or two of their own residences. Between them, they built 22,500 homes in the area.[34] Each city also had its big operators: developers or subdividers who had extended their operations to financing and building. They became a key part of their city's political-economic growth machine. One such was Samuel Eberly Gross, who had styled himself as "the World's Greatest Real Estate Promoter" with "the Largest Real Estate Business in the World" in Chicago even before the streetcar boom. When the streetcar era arrived, Gross promptly became part of Chicago's growth machine, planning streetcar lines and developing the surrounding properties.[35]

A great deal of the development in streetcar suburbs fell well short of the ideals and plans for architecture and urban design that intellectuals and designers had developed in previous decades. Unregulated, speculative development in fiercely competitive markets led to monotonous layouts, shoddy workmanship, incomplete infrastructure, and perfunctory landscaping. For the upper-middle class at the leading edges of suburbanization, encroachment by substandard subdivisions, the rapid deterioration of once-fashionable developments, and the possibility of the arrival of "undesirable" people and activities were deep-seated and increasingly well-founded fears. Robert Fogelsong, in *Bourgeois Nightmares*, shows how subdividers turned increasingly to the imposition of restrictive covenants, or deed restrictions, in their attempts to impose exclusivity and stability on their products.[36] This led to the emergence of a new type of subdivider: the "community builder," a full-fledged suburban housing developer, not only planning and improving large lots of land but also building the housing on the lots and selling the completed package to the homebuyer. Marc Weiss, in his book on *The Rise of the Community Builders*, shows how these new developers helped introduce and implement key concepts of planning and urban design to American suburban landscapes. "The classification and design of major and minor streets, the superblock and cul-de-sac, planting strips and rolling topography, arrangement of the house on the lot, lot size and shape, set-back lines and lot coverage restrictions, planned separation and relation of uses, design and placement of parks and recreational amenities, ornamentation, easements, underground utilities, and numerous other physical features were first introduced by private developers and later adopted as rules and principles by public planning agencies."[37] Meanwhile, less affluent households aspiring to become homeowners had to find large down payments, and only short-term mortgages were available. As Matt Edel, Elliot Sclar, and Daniel Luria demonstrated in *Shaky Palaces*, many lower-middle-class households found themselves struggling "up the down escalator," entranced by dreams of economic security, saddled with debt, and confused by a false sense of social mobility.[38] Sustaining the enchantment of bourgeois utopia in the face of these realities not only required a great deal of promotional advertising but also the professionalization of real estate brokerage.

The boom in streetcar suburbs saw real estate brokerage transform from a loosely and locally regulated activity, open to any unscrupulous operative, to a nationally organized occupation with significant influence on American housing policy and the powerful notion of the "American Dream." The term was first coined in 1931 by James Truslow Adams in his book *The Epic of America*.[39] The product of Depression-era politics, the original notion of the American Dream built on the idea of American exceptionalism, stressing individual freedom, especially the possibility of dramatic upward social mobility through ingenuity and hard work, with the promise that successive generations would enjoy steadily improving economic and social conditions. It did not take long, though, for the ideal of home ownership to be grafted on to the notion. As Jeffrey Hornstein notes in his book *Nation of Realtors*, "The conception of real estate brokerage as an occupation-*cum*-profession depended upon the existence of 'home' as an intellectual and cultural object. . . . Thanks in large measure to real estate brokers' cultural and political work, the single-family home on a quarter-acre lot in a low-density suburban development became the 'American Dream,' and the vast majority of Americans bought into it."[40] The National Association of Real Estate Exchanges (subsequently renamed as the National Association of Real Estate Boards, and renamed again in 1973 as the National Association of Realtors) was established in 1908. Collectively, "realtors"—the brand name adopted and registered by the national association in 1916[41]—systematically developed and deployed strategies to sell houses that helped to reify this notion of home and neighborhood as the "normal" desideratum for a "middle class" American family.

From about 1915 through the 1920s, realtors collaborated with various government agencies and civic groups to promote single-family home ownership. Their Own-Your-Own-Home campaign sought to reinforce the idea of the home as a privileged consumer durable, worth sacrificing and going into debt for. This mission to make America into a land of universal homeownership "was merely the culmination of a long republican tradition linking civic virtue to property ownership. The American republic would be able to save itself from the degenerative ravages of historical time, class struggle, and urban corruption by providing all citizens with a home of their own in healthful, natural surroundings: American civilization would develop in space rather than time."[42]

Local real estate boards regularized the rules for participation in home ownership by creating standardized paperwork and uniform commission schedules. By the 1920s, the national association had promulgated model real estate broker licensing and zoning laws drawn up by member boards, which municipal and state governments often adopted verbatim. It had also endorsed standard courses in land economics and real estate appraisal and practice. When the housing market crashed at the onset of the Depression, the National Association of Real Estate Boards was in a position to work closely with President Herbert Hoover's White House Conference on Home Building and Home Ownership. In doing so, it secured support for a reduction of taxes on real estate and endorsement for a federal mortgage discount bank to facilitate long-term mortgages. The conference also endorsed a policy of preference for large-scale suburban development in order to take advantage of economies of scale, giving favored status to large, planned suburban developments for securing government-backed financing. These became key elements in the ambitious programs of the Roosevelt administration's New Deal. Thus, by the mid-1930s, realtors' political and cultural work had "created national housing policy that privileged suburban sprawl, middle-class privatization of social space, and racial segregation."[43]

Democratic Utopias

The arrival of mortgage insurance for lenders, together with the increasing affordability of automobiles and of the application of production-line manufacturing techniques to large-scale housing subdivisions, laid the foundations for

contemporary suburbia. Frank Lloyd Wright, an architect who fancied himself
as a visionary intellectual, could see it coming. Wright, ever the iconoclast, decided
to be an architect who hated cities. He was also anxious to position himself as a
visionary with a distinctively American flavor. His vision was for a "Usonian" (a
word-play on U.S. own) future. In contrast to the dominant Modernism of the
day, and in direct contradiction of the rationale of the high-rise Ville Radieuse of
Le Corbusier (another architect who fancied himself as an intellectual visionary), Wright argued for a low-density, low-rise pattern of settlement in his idealized Broadacre City. Drawing on the individualism and naturalism of Jefferson,
Thoreau, and Emerson, Wright took a stance that gave primacy to individual freedom rather than to the Modernists' emphasis on social democracy. Single-family
homes on one-acre lots, he argued, provided the only way to guarantee the individual freedom that was the birthright of Americans; Broadacre City would be
the ultimate expression of a truly democratic society.[44] Further, it would be
healthful, aesthetically pleasing, and morally and culturally uplifting. The inac-
cessibility inherent to the large lots and low densities of Broadacre City was to
be conquered by a network of landscaped parkways and freeways, with the focal
point of semi-rural neighborhoods being provided by huge gas stations, architectural centerpieces that would double up as cafeterias and mini-marts. Unfortunately Wright, like most would-be visionary architects, did not really understand
cities and their complex, recursive interdependencies. Like his predecessors, his
contemporaries, and his successors, he was able to see no further than a prescriptive and deterministic relationship between urban design and individual and
social well-being. This will be, rather tiresomely, a recurring theme.

But although Wright's rationale was flawed, his inspiration—the automobile and the wide-open spaces of rural America—proved to be the dominant
ingredients of the unfolding suburban landscape. The impetus came not from
any prescriptive designs or compelling vision but from the rather more prosaic
combination of federal policy, judicial ruling, political economy, and the increasing affordability and efficiency of automobiles. The market failures that had triggered the Depression undermined the legitimacy of classical, *laissez-faire*
liberalism and led to its eclipse by an egalitarian liberalism that relied upon the
state to manage economic development and soften the unwanted side effects of
free-market capitalism. Central to federal policy in the 1930s was federal mortgage insurance, part of a broad package of Keynesian macroeconomic management
introduced by the Roosevelt administration in response to the overaccumulation
crisis of the Depression.[45] Roosevelt's Federal Housing Administration (FHA),
established in 1934, played a key role in stimulating the labor-intensive construction industry by stabilizing the mortgage market and facilitating sound
home financing on reasonable terms. The Federal Housing Administration also
established and enforced minimum standards for the housing financed by its
guaranteed loans, thus helping to eliminate shoddy suburban construction. The
Federal Housing Administration was by no means singularly progressive, however. In its 1939 *Underwriting Manual*, the Federal Housing Administration

openly recommended that subdivision developers use restrictive covenants to prevent the sale of homes to minorities. Mortgage redlining, which designated certain sections of an urban area as unsuitable for Federal Housing Administration-insured mortgages, was common practice. These efforts were intended to reduce the risk that homeowners would default on their mortgages. It was not until 1949 that discriminatory restrictive covenants were declared unconstitutional. By then the Federal Housing Administration had formulated the nation's suburban planning template, including subdivision design standards, mortgage redlining practices (which lasted into the 1980s and arguably continue through various stealthy tactics), and a bias toward detached single-family owner-occupied housing.[46]

The immediate effect of Keynesian policy was to reignite suburban growth, creating a "spatial fix" to the overaccumulation crisis.[47] Whereas housing starts had fallen to just over 90,000 in 1933, the number of new homes started in 1937 was 332,000, and in 1941 it was 619,000. Optimistic New Deal administrators saw a further opportunity: to *plan* suburban development, democratizing the suburbs by drawing people from redeveloped central cities. Rexford Guy Tugwell, appointed by Roosevelt as head of the Resettlement Administration, envisaged some 3,000 "greenbelt" cities that would contain government-sponsored low-cost housing, and promptly drew up a list of twenty-five cities on which to start. Funds were allocated for only eight, however, and Congress, under strong pressure from the private development industry, whittled this number down to five. Two of the five were blocked by local legal action. The remaining three were built: Greendale, southwest of Milwaukee; Greenhills, near Cincinnati; and Greenbelt, just north of Washington, D.C. But in 1938 the Resettlement Administration was abolished and after World War II all three were sold off to nonprofit corporations, to be swallowed up soon afterward in the continuing sprawl of automobile suburbs.

It was the automobile, together with a broad raft of federal policies, that was to democratize the suburbs and bring the fulfillment of the American Dream to a broader cross-section of society.[48] Automobiles, however, are of little use without good roads. The first push for good roads had come as early as the 1890s, from bicyclists, the post office, and farmers' Granges—associations that wanted to promote better access to markets. This was consolidated a decade later when the automobile, oil, rubber, and construction industries were joined by the Good Roads Association, a confederation of urban merchants and industrialists who believed that better highways would improve business.[49] Their campaigns resulted in the 1916 Federal Aid Roads Act, which required every state, as a condition of federal aid, to establish a state highway department to plan, build, and maintain interurban highways. A second act in 1921 provided additional funds with the objective of integrating the long-distance road network. In the 1940s and 1950s, lobbying for federal support for highway construction was taken up by an informal but very powerful group that came to be known as the Road Gang (or, alternatively, the Highwaymen) and included automotive manufacturers, auto clubs, oil and tire companies, and highway engineers.[50]

The enchantment of the suburbs to middle-class whites was meanwhile reinforced by the U.S. Supreme Court in its landmark case on zoning law: *Village of Euclid, Ohio v. Ambler Realty Co.* (1926). Ruling in favor of the municipality's right to prevent a property owner from using land for purposes other than for which it had been zoned, the Court established the power of local governments to "abate a nuisance." The latter, the Court ruled, could be defined very broadly to include anything affecting the general welfare of a residential area. As a result, zoning promptly came to be used to exclude not only undesirable land uses from residential areas but also (by establishing large—and therefore expensive—minimum lot or dwelling sizes, for example) undesirable people. Growing suburban areas rushed to become incorporated as municipalities in order to be able to control the pace and nature of growth.

Speculative developers, reassured by the stability conferred on the land market by the zoning maps established by these jurisdictions, were emboldened to lay out larger and larger subdivisions, copying, as far as possible, Henry Ford's techniques of mass production for mass consumption. This required the pursuit of economies of scale, the standardization of products, the perfection of prefabrication technology, and the rationalization of street layouts. Many developers dispensed with sidewalks altogether, partly to save money, partly to emphasize the exclusivity of the neighborhoods to the automobile-owning classes, and partly to lend an arcadian feel. The consequences for suburban development were far-reaching. The Fordist period was characterized by a spectacular increase in housing construction. From the low point of just over 90,000 housing starts in 1933 and a pre-war average of about 350,000 starts per year, construction jumped to almost two million by 1950. In these new suburbs, lot sizes were larger and densities were lower. The average size of a building lot rose from about 3,000 square feet in streetcar suburbs to about 5,000 square feet in automobile suburbs; residential densities fell from about 20,000 people per square mile in streetcar suburbs to between 5,000 and 8,000 per square mile in early postwar automobile suburbs.

Taking advantage of pattern-book designs and balloon-framing techniques, developers began to encircle every city of any significance with huge, sprawling subdivisions. One of the largest was the Lakewood Park complex, built to accommodate more than 100,000 people on sixteen square miles of brushland south of Los Angeles. Without doubt the most famous was the original Levittown on Long Island, begun in 1947 by Abraham Levitt and his sons William and Alfred. They were the first large-scale developers to apply a highly rationalized, assembly-line approach to residential development. Combined with the use of innovative materials, new tools, standardized designs, low prices (the original Levittown Cape Cod design sold for $100 down and $57 per month), and, finally, slick marketing, this approach unrolled more than 17,000 homes onto the Long Island suburban fringe in short order.

After 1945 there began a second spurt of growth in automobile ownership. There was a backlog of unfulfilled demand for housing from the Depression and war years, combined with the postwar baby boom. During the war there had been a moratorium on new construction, so that by 1945 there was an accumulated backlog of between three and four million dwellings. In 1944 the Servicemen's Readjustment Act (the "GI Bill") created the Veterans Administration, one of the major goals of which was to facilitate home ownership for returning veterans. It did so through a program of mortgage insurance along the lines of the Federal Housing Administration, whose own lending powers were massively increased under the terms of the 1949 Housing Act. There was also another significant phase of road building as a result of the Federal Aid Highway Act (1956), which authorized 41,000 miles of limited-access highway. Every major city was to be linked into the system, with circumferential beltways that made outlying locations more accessible, particularly where there were intersections with major radial interstate spokes. Meanwhile, the number of automobiles on the roads jumped from just under twenty-six million in 1945 to more than fifty-two million in 1955 and ninety-seven million by 1972.

The result was a dramatic spurt in suburban growth. David Harvey has argued persuasively that it was part of an overall strategy to create and maintain a long-term cycle of growth, an extended "spatial fix" underwritten by massive outlays for defense and freeway construction and subsidies for the suburban real estate sector.[51] The combined effect was to stimulate jobs not only in defense industries and construction but also in automotive industries and consumer durables. From a demand-side perspective, it was also a continuation of the fast-paced suburbanization of the 1920s, which had been interrupted by the depression of the 1930s and World War II. The 1950s was the decade of the greatest-ever growth in suburban population. While central cities in the United States grew by six million people (11.6 percent), suburban counties added nineteen million people (45.9 percent). Within almost every metropolitan area, the ring of suburban counties grew much faster than the central city (or cities). This spurt coincided with a dramatic postwar increase in prosperity and the consequent rise of consumerism. Between 1948 and 1973 the economy grew at unprecedented rates. The Gross National Product increased fivefold, median income more than doubled (in constant dollars), and home ownership rose by 50 percent.

Historian Lizabeth Cohen has traced the development of a "consumers' republic" in the United States in this era: a society based on mass consumption of automobiles, houses, and manufactured household goods, all celebrated by the new medium of television.[52] This was the era of the "Sitcom Suburb," a democratic utopia of ranch and split-level homes, all stocked with dozens of small appliances (Figure 2.1). Thanks to *Euclid v. Ambler*, Sitcom Suburbs were founded on local government zoning regulations that prohibited apartments, duplexes, small houses, or small lots as well as stores and offices. Federal intervention also contributed significantly to the creation of standardized suburban settings dominated by detached single-family homes occupied by white families. With homeownership its principal objective, the Federal Housing Administration was clearly biased from its inception toward single-family detached and owner-occupied housing. To assist local governments with planning for single-family detached

FIGURE 2-1. Sitcom Suburbia: A new democratic utopia of family-oriented settings that became the foundation for a new phase of material consumption. (© H. Armstrong Roberts/Corbis)

homes, the Federal Housing Administration recommended standardized subdivision design practices that became a template for suburban subdivisions nationwide when Congress passed the landmark Section 701 planning grant program as part of the 1954 Federal Housing Act. In this controlled setting, life imitated television, where "Moms in high heels and dresses heated frozen dinners in commercials that seemed like extensions of family-oriented, prime-time programming. Thousands of television commercials and print ads used the model house as the setting for all sorts of goods from detergents to diapers, dishwashers to Dodge cars."[53] This was the pre-*Cosby* and *Jamie Foxx Show* sitcom era, however. Sitcom Suburbs were rolled out within local government zoning regulations that were founded on the exclusionary potential of the *Euclid v. Ambler* decision and then marketed through ongoing practices of mortgage redlining and discrimination.

Rising incomes, broadening educational opportunities, increasing levels of home ownership, and ever-expanding choices of consumer goods allowed most Americans to reimagine their country and what it meant to be an American. Within a decade or two of the end of World War II, Americans had developed a distinctive way of life and a new social and spatial order known as suburbia. As industrial cities declined, sunbelt cities and suburbia displaced them as the cradles of national personality. The suburbs' new centrality to American identity was reinforced during the Cold War as the United States showcased its democratic utopia of suburban lifestyles and consumer culture by way of contrast with the Soviet Union's regimented lifestyles and modest levels of living. "Awash in consumer goods, enjoying nearly full employment, and blessed with high wages, the daily life of the 'average' American became a model for people around the globe. Suburban life anchored a standard of living commensurate with the nation's status as the leader of the 'free world' and established the country's economy and form of government as the best hope for affluence, democracy, and world peace."[54] This new narrative also enabled Americans to distinguish themselves from the old-world culture associated with European cities, simultaneously adding another dimension to the idea of American exceptionalism.

Suburbia not only became the locus for the realization of the American Dream but also the crucible for what David Brooks calls a "Paradise Spell" of relentless individual aspiration and restless consumption: "Born in abundance, inspired by opportunity, nurtured in imagination, spiritualized by a sense of God's blessing and call, and realized in ordinary life day by day, this Paradise Spell is the controlling ideology of American life."[55] This, of course, fits squarely with the notion of "romantic capitalism" described in Chapter 1: a consumer economy driven by dreams and fantasies.[56] But, under this Spell, the individual pursuit of the Dream has led inevitably to collective disappointment. John Archer puts it this way: "In postwar decades the 'dream' progressively became available to an increasingly democratic public, . . . waves of white- and blue-collar workers who could afford mass-produced houses in tract developments. The dream likewise proliferated through ever more affordable mass products and standardized materials. A consequence of such affordability was that products were designed and marketed according to increasingly reductive stereotypes. And this points to an implicit contradiction embedded in the democratization of the dream: the ideal of a personalized dream individually defined and achieved is inconsistent with its material realization in stereotyped forms and mass-produced materials."[57] Standardization begets indifference; rationalization leads to disenchantment. By the early 1970s, the Sitcom Suburbs of the 1950s and early 1960s were tinged with an aura of failed ambition, and disenchantment with suburbia had become the conventional wisdom. In spite of the fact that there were few practical alternatives to the "sprawl machine,"[58] suburbia had acquired a charge sheet that listed not only the bland standardization and rationalization of "placeless" subdivisions but also environmental degradation, social isolation, and malaise.

In the mid-1970s, the economic system-shock triggered by the Organization of the Petroleum Exporting Countries (OPEC) cartel's fourfold increase in the price of crude oil (in 1973) not only dampened consumers' ability and enthusiasm to pay for suburban lifestyles but also set in motion a shift away from the egalitarian liberalism that had dominated public policy since the 1930s. Just as the idea of market failures had been a powerful notion in the ideological shift from classical liberalism to egalitarian liberalism in the 1930s, so the idea of government failures became a powerful notion in undermining egalitarian liberalism (and especially the Keynesian welfare state) in the mid- to late 1970s. Governments, the argument ran, were inefficient, bloated with bureaucracy, prone to over-regulation that stifles economic development, and committed to social and environmental policies that are an impediment to international competitiveness. As a result, egalitarian liberalism was eclipsed by neoliberalism, a selective return to the ideas of classical liberalism.[59] The Reagan administration of the 1980s dismantled much of the Keynesian welfare state, deregulated industry (notably including the mortgage finance and real estate sectors), ushered in an era of public-private cooperation in place-making and economic development, and rekindled libertarian ideas about the primacy of private property rights. As we shall see, these were to become powerful elements in redirecting urban change. Meanwhile, the federal government withdrew much of its support for the "spatial fix" that had supported a democratic utopia of sitcom suburbs: road building required a level of taxation that was anathema to neoliberalism, while the end of the Cold War led to a retrenchment in defense spending.

Disenchantment: Sprawl and Placelessness

The intellectual critique of suburbia had in fact begun with the first large-scale developments. Lewis Mumford began to chronicle the transition from utopia to subtopia as early as the 1920s, borrowing Patrick Geddes's memorable phrase, "more and more of worse and worse."[60] After World War II, the critique intensified. A satirical portrayal of the lives of "John and Mary Drone" in the suburbs around Washington, D.C., in the mid-1950s was caustic:

> For literally nothing down . . . you too . . . can find a box of your own in one of the fresh-air slums we're building around the edge of America's cities . . . inhabited by people whose age, income, number of children, problems, habits, conversation, dress, possessions and perhaps even blood type are also precisely like yours. . . . [They are] developments conceived in error, nurtured by greed, corroding everything they touch. They destroy established cities and trade patterns, pose dangerous problems for the areas they invade, and actually drive mad myriads of housewives shut up in them.[61]

The editors of *Fortune* magazine sponsored a conference that staged a bitter attack on postwar suburban development, while the influential Mumford, in his summative work on the history of cities, let his intellectual elitism show through in an intensified critique of democratized suburbia:

> A multitude of uniform, unidentifiable houses, lined up inflexibly, at uniform distances, on uniform roads, in a treeless communal waste, inhabited by people of the same class, the same income, the same age group, witnessing the same television performances, eating the same tasteless pre-fabricated foods, from the same freezers, conforming in every outward and inward respect to a common mold. . . . The ultimate effect of the

suburban escape in our own time is, ironically, a low-grade uniform environment from which escape seems impossible.[62]

Borrowing freely from Geddes's biological metaphors, Mumford described suburban sprawl as cancerous, with "social chromosomes and cells" running riot in "an overgrowth of formless new tissue. . . . The city has absorbed villages and little towns, reducing them to place names . . . as one moves away from the center, the urban growth becomes ever more aimless and discontinuous, more diffuse and unfocused."[63]

Meanwhile, the intellectual basis for *planned* development of any kind took a severe blow with the publication and popularity of Jane Jacobs's *The Death and Life of Great American Cities*. Jacobs, an arch enemy of Mumford, was a champion of urbanism and the rich diversity of central city neighborhoods. The problem with every kind of planned development, from garden suburbs to the City Beautiful, and from developer subdivisions to public housing projects, she charged, was that architects, planners, and developers impose their designs for order and efficiency without understanding or respecting the "close-grained diversity," the unexpected, unplanned, and incongruous that is the true glory of city life.[64]

In popular music, Pete Seeger sang about little boxes made of "ticky-tacky" that all looked the same; Joni Mitchell complained that "they paved paradise and put up a parking lot"; and Chrissie Hynde recounted that:

> I went back to Ohio
> But my pretty countryside
> Had been paved down the middle
> By a government that had no pride
> The farms of Ohio
> Had been replaced by shopping malls
> And muzak filled the air
> From Seneca to Cuyahoga Falls[65]

The shocking rate and extent of low-density suburban development was behind a lot of the popular critique of suburbia. By the late 1960s, "sprawl" had become a synonym for suburbia. It did not take long for academics, architects, planners, and social commentators to document the nexus of negative attributes of sprawl.[66] The unplanned, ad hoc nature of most suburban development, it was pointed out, destroys millions of acres of wildlife habitat and agricultural land every year. Rationalized, standardized, and tightly zoned suburban developments result in neighborhoods that lack visual, demographic, and social diversity. The economics of private subdivision lead to a lack of public open space, urban infrastructure, and civic amenities. The low densities inherent to single-family suburban development result in increased traffic, long commutes, and a chronic dependence on automobiles. The environmental costs of automobile dependency include air pollution—and in particular the generation of millions of tons of greenhouse gases from suburban commuters—and polluted run-off from the roads and parking lots that constitute a third or more of suburban watersheds. The automobile-dependent lifestyles associated with sprawl, meanwhile, lead to increases in rates of asthma, lung cancer, and heart problems. Stress resulting from commuting leads to adverse effects on marriages and family life. The fragmented and balkanized nature of American local government means that sprawl also intensifies intra-metropolitan fiscal disparities: outlying communities have a larger tax base and fewer social service needs to finance in comparison with central cities. This structured inequality exacerbates the exclusionary nature of suburbia and, from a planning perspective, is simply inefficient. As historian Jon Teaford succinctly puts it, "Sexual intercourse, connubial affection, motherly devotion, atmospheric purity, flora and fauna, civic loyalty, and individual happiness all seemed to be victims of the relentless sprawl of the edgeless city."[67]

By the late 1980s disenchantment with sprawl—and therefore with suburbia—was being actively mobilized by a coalition of nonprofit institutions that included the Sierra Club, the Natural Resources Defense Council, the Rural Heritage program of the National Trust for Historic Preservation, and the American Farmland Trust. There have been few apologists for sprawl in the face of this popular conventional wisdom. Architectural historian Robert Bruegmann has attempted a revisionist/apologist history of sprawl, arguing that sprawl is a logical consequence of economic growth and the democratization of society, providing millions of people with the kinds of mobility, privacy, and choice that were once the prerogatives of the rich and powerful. Similarly, Deyan Sudjic, in *The 100-Mile City*, argues that the development of metropolitan regions, while naturally reflecting market forces and representing the outcome of public and private planning, also expresses the desires and preferences of the broad mass of people. Joel Kotkin, a popular speaker on the business circuit, has persistently argued that most people seem to like living in the suburbs: so why all the fuss? Libertarian think-tanks like the Reason Foundation and the Heritage Foundation have made efforts to counter anti-sprawl arguments on the principle of freedom of markets and individual choice. The title of Wendell Cox's book on the subject gets straight to the point: *War on the Dream: How Anti-Sprawl Policy Threatens the Quality of Life*.[68] These neoliberal interpretations appeal to the individualistic republicanism that derives from the American Renaissance. What they overlook, as Alex Krieger points out, is that "the benefits of sprawl—for example, more housing for less cost with higher eventual appreciation—still tend to accrue to Americans individually, while sprawl's cost in infrastructure building, energy generation, and pollution mitigation tends to be borne by society overall."[69]

Post-Utopian Suburbia
The utopian aura of suburbia that was the legacy of the nineteenth century was comprehensively undermined by classic social science monographs of the 1950s and 1960s such as *The Eclipse of Community, The Lonely Crowd, The Organization Man, The Levittowners, Middletown*, and *Working Class Suburb*.[70]

Although differing in their focus and emphasis, these studies established the image of the suburbs as settings of loose-knit, secondary ties where lifestyles were focused squarely on the nuclear family's pursuit of money, status, and consumer durables and the privacy in which to enjoy them. Suburban neighborhoods were portrayed as "communities of limited liability"—just one of a series of social networks in which people might choose to participate. Within the suburbs themselves, conformity, shallowness, isolation, and the separation of the public and private spheres of life were hallmarks. A century after the publication of *Walden*, it was the mass of residents of suburbia who were evidently leading lives of quiet desperation.

These themes still dominate the literature, both academic and popular, fiction and nonfiction. Robert Putnam's briefcase best seller *Bowling Alone*, for example, focuses on the growing separation of home and work, the increasing segregation and homogeneity of suburbia, and the attenuation of civic engagement. In a more populist vein, David Brooks, in his essay on "Patio Man and the Sprawl People," has lampooned the way that contemporary suburbia is dominated by a politics of identity in which consumption has become the central locus of competition.[71] It is the placelessness of suburbia, however, that has become one of its most emblematic attributes.

The Geography of Nowhere

It was another briefcase best seller, *The Geography of Nowhere*, by James Howard Kunstler, that captured popular disenchantment with the placelessness of suburbia: the "Nowhere" of the book's title. Kunstler portrayed suburbia as a cartoon landscape of tract houses, car-clogged highways, parking lots, strip malls, and franchise food, with no sense of place. The *genius loci* of towns and cities has been eclipsed by the *genius loco* of suburbia. Geographer Edward Relph was in fact the first to identify and analyze the placelessness of North American urban settings, connecting it to the rationalization and standardization of post–World War II subdivisions. More recently, architect Rem Koolhaas has critiqued the cloned subdivisions and hallucinogenic normality of "Generica," the product of a building logic driven by short-term efficiencies: agility, turnover, and scale. Kunstler, lamenting the decline of "real" urban settings, points out that fewer and fewer Americans have any experience of good cities, or indeed in any town or city at all. The inference is that Americans no longer know enough as consumers of housing to be discriminating buyers, though, as Witold Rybcznski notes, the architecture profession and professional schools turned away from the design of suburbs and suburban housing after the 1930s, leaving the field to developers' pattern books and builders' impulses.[72]

But sense of place is not simply a matter of the aesthetics of the built environment. It is always *socially* constructed, and a fundamental element in the social construction of place is the existential imperative for people to define themselves in relation to the material world. The roots of this idea are to be found in the philosophy of Martin Heidegger, who contended that men and women originate in an alienated condition and define themselves, among other ways, through their sociospatial environment. People's "creation" of space provides them with roots—their homes and localities becoming biographies of that creation. Central to Heidegger's philosophy is the notion of "dwelling": the basic capacity to achieve a form of spiritual unity between humans and the material world. Through repeated experience and complex associations, our capacity for dwelling allows us to construct places, to give them meanings that are deepened and qualified over time with multiple nuances. Another crucial concept here is that of the lifeworld, the taken-for-granted pattern and context for everyday living through which people conduct their day-to-day lives without having to make it an object of conscious attention. People's experience of everyday routines in familiar settings leads reflexively to a pool of shared meanings. Neighbors become familiar with one another's vocabulary, speech patterns, dress codes, gestures, and humor, and with shared experiences of the physical environment such as streets, markets, and parks. Often this carries over into people's attitudes and feelings about themselves and their locality and to the symbolism they attach to that place. When this happens, the result is a collective and self-conscious "structure of feeling": the affective frame of reference generated among people as a result of the experiences and memories that they associate with a particular place.[73]

Heidegger anticipated the effects of rationalism, mass production, standardization, and mass values on people's capacity for "dwelling" and the social construction of place. The inevitable result, he suggested, is that the "authenticity" of place is subverted. Neighborhoods become inauthentic and placeless, a process that is, ironically, reinforced as people seek authenticity through professionally designed and commercially constructed spaces and places whose invented traditions, sanitized and simplified symbolism, and commercialized heritage all make for convergence rather than spatial identity. This particular dialectic of enchantment and disenchantment is taken up in detail in chapter 6.

The basis of both individual lifeworlds and the collective structure of feeling is intersubjectivity: shared meanings that are derived from the lived experience of everyday practice. An important part of the basis for intersubjectivity is the routinization of individual and social practice in time and space. A positive and distinctive sense of place stems in large part from routine encounters and shared experiences that make for intersubjectivity. This requires plenty of opportunities for informal, casual meetings and gossip; friendly bars and pubs and a variety of settings in which to purchase and/or consume food; street markets; a variety of comfortable places to sit, wait, and people-watch; a sense of ease with changing seasons; and, above all, a sense of belonging, affection, hospitality, vitality, and historical and cultural continuity: not the attributes that anyone associates with American suburbia.

In terms of Anthony Giddens's structuration theory (see chapter 1), the placelessness of suburbia is central to an understanding of a political economy in which time-space distanciation has "emptied out" time and space, with the result that people have become "disembedded" from their localities. This can produce feelings of being strangely out of place while in everyday settings—*unheimlich*, in Freud's terms.

Suburban Malaise

Sense of place and authenticity notwithstanding, the popular image of suburbia has deteriorated as democratization and sheer weight of numbers have imported the full spectrum of maladies that afflict American society. Suburbia's own particular social pathology is the collective ritual of frustration and misery that is commuting. Sprawl has increased both the trip length and the travel time of commuters. According to American Community Survey data, over 220 million adults average an hour and a half a day in their cars. "Extreme" commuters—3.4 million of them in 2005—spend up to a month of their lives each year traveling a minimum of an hour and a half to work and back.[74] Their ranks have increased by 95 percent since 1990, making the Blue Ridge Mountains a bedroom community of Washington, D.C., New Hampshire an exurb of Boston, and Modesto, California, an outpost of Silicon Valley. New York and Baltimore have the highest percentage of people with "extreme" commutes: 5.6 percent of their commuters spent ninety or more minutes getting to work. People with extreme commutes were also heavily concentrated in Newark, New Jersey (5.2 percent); Riverside, California (5.0 percent); Los Angeles (3.0 percent); Philadelphia (2.9 percent); and Chicago (2.5 percent). In the Washington area, nearly 60 percent of commuters say they get tangled up in traffic jams at least once a week, and 28 percent say they encounter serious tie-ups every day. In 1982, the average Washington area commuter spent twenty-one hours stuck in traffic. That more than doubled to forty-eight hours by 1992 and jumped to sixty-seven in 2002.[75]

Seduced into making long commutes by the prospect of a large house at an affordable price, many suburbanites have to leave home at six A.M. or earlier, only to arrive back in time for a late dinner. Extreme commuters spend twice as much time driving as they do with their families. Social connections also suffer. According to Robert Putnam, in *Bowling Alone*, social connections fall away by ten percent for every ten minutes of commuting time. In the medical literature, commuting is associated with the etiology of raised blood pressure and musculoskeletal disorders. It is also associated with an increased incidence of lateness, absenteeism, adverse effects on cognitive performance, and increased hostility. Road rage is the most vivid manifestation of this hostility. Gregory Squires recounts a particularly nasty example: "One fall afternoon in 1999, two soccer moms played cat-and-mouse games during the homebound rush hour on the increasingly congested Interstate 65 outside of Birmingham, Alabama. After tailgating, lane changing, and brake slamming over four miles of expressway, they came to a red light. Gena Foster jumped out of her sport-utility vehicle and approached Shirley Henson in her Toyota 4Runner. Henson grabbed her .38 caliber revolver, lowered her window, and shot Foster to death."[76] The American Automobile Association's Foundation for Traffic Safety reported an increase of road rage incidents of nearly 60 percent between 1991 and 1996, resulting in 218 deaths and 12,610 injuries—mostly accidental—during the period. Aggressive driving causes around 28,000 deaths a year, and traffic accidents in general are responsible for twice as many fatalities as violent crime, so that the suburbs are, on the whole, much more dangerous than central cities. In the Washington, D.C., metropolitan area for example, combined traffic fatalities and homicides by strangers are significantly higher in the suburbanized commuter counties of Clark, Fauquier, Frederick, and Stafford than in the District of Columbia—despite the District having one of the nation's top ten homicide rates.[77]

But in addition to the pathologies of commuting, suburbia now suffers from its share of "urban" malaise. Although central city crime rates are typically twice those of the suburbs, suburban crime is increasing at a much faster rate; and often it is increasing faster than the rate of suburban population growth. Burglary, historically a night-time, inner-city crime, has become a weekday suburban crime because that is when and where homes are empty and vulnerable. The opportunities for suburban crime have increased: more bank branches and automated teller machines, more copy shops and drugstores, and more teens in the market for drugs. Targets for property crime in the suburbs have meanwhile become more accessible, thanks to new and wider highways that make it easier for criminals to move quickly in and out of an area. Boredom and ennui, meanwhile, have turned some tracts of suburbia into "teenage wastelands," a social ecology that features burnouts, hangouts, dropouts, goths, cliques, and gangs.[78]

Violent crime has also become part of suburban social ecology. In the Washington area, for example, the combined numbers of homicides, rapes, robberies and serious assaults have risen even faster than the population in thirteen of sixteen major suburbs over the past twenty-five years.[79] Washington suburbs were also the setting for an episode of the singularly American crime of recreational sniper murders. John Allen Muhammad and Lee Boyd Malvo terrorized the region, shooting sixteen people—and killing ten of them—from a concealed position in the trunk of their car over the course of forty-seven days in September and October 2002. High school shootings have also become emblematic of violent crime in suburbia. Between 1996 and 2005 there were twenty-six separate incidents of high school shootings in the United States, all of them in suburban settings. The most notorious was in Littleton, Colorado, in 1999, where Eric Harris, eighteen, and Dylan Klebold, seventeen, had plotted for a year to kill at least 500 and blow up their school, Columbine High. They killed twelve of their fellow students and one teacher and wounded twenty-three others before turning their guns on themselves. Suburbia in this context is the new heart of darkness, a city of dreadful day rather than dreadful night, with terrifying crimes carried out in the bleak boredom of bland subdivisions.

These are certainly extreme examples, but they are rooted in the social ecology of suburban settings within which are embedded almost every facet and element of contemporary American society. In this situation, it is hard to sustain utopian ideals or even commercial enchantment. Yet, as the evolution of metropolitan regions unfolds, the leading edges of metroburbia are reframed and reenchanted, featuring the developers' utopias and degenerate utopias of the New Metropolis.

CHAPTER 3

Metroburbia and the Anatomy of the New Metropolis

The sheer scale and extent of contemporary development in America has challenged our lexicon of urbanization. It also severely challenges the models of urban form and structure that for so long have been the staples of urban studies. The suburbs now not only contain the largest fraction of America's households and population but also a significant fraction of America's industry, commercial office space, retailing, recreational facilities, and tourist attractions. Traditional "laws" of urbanization are being repealed by massive changes in real estate investment, in tandem with equally significant changes in the structure and functional organization of metropolitan regions. The simultaneous shift from traditional national agencies of governance and development upward to global or supranational institutions and downward to nonprofit organizations and local actors—"glocalization"—has unleashed the dynamics of urbanization to "wrestle free from the cocoon in which the managed capitalism and planned modernity of the postwar era had tried to contain them."[1] New rounds of economic restructuring, advances in digital telecommunications technologies, demographic shifts, and neoliberal policies are giving rise to new urban, suburban, and exurban landscapes. Urban regions are being stretched and reshaped to accommodate increasingly complex and extensive patterns of interdependency, while the political economy of metropolitan America is being reshaped in response to socioeconomic realignments and cultural shifts.

Until the middle of the twentieth century, urban and metropolitan form could safely be conceptualized in terms of the outcomes of processes of competition for land and of ecological processes of congregation and segregation—all pivoting tightly around a dominant central business district and transportation hub (Figure 3.1a). During the middle decades of the twentieth century, however, American metropolises were unbound by the combination of increased automobility and the blossoming of egalitarian liberalism in the form of massive federal outlays on highway construction and mortgage insurance. The resulting massive spurt of city building produced a dispersed spatial structure and the emergence

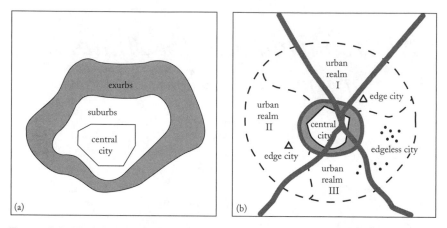

FIGURE 3-1. Evolving twentieth-century metropolitan form. (Paul L. Knox and Robert E. Lang, The Metropolitan Institute, Virginia Tech)

of "urban realms"—semi-autonomous sub-regions that resulted from a polycentric metropolitan structure (Figure 3.1b) that displaced the traditional core-periphery relationship between central cities or principal cities[2] and their suburbs. Initially, the shift to an expanded polycentric metropolis was most pronounced in the northeastern United States, and Jean Gottmann captured the moment in 1961 with his conceptualization of "megalopolis"—his term for the highly urbanized region that stretched between Boston and Washington, D.C. Pierce Lewis subsequently coined the term "galactic metropolis" to describe the disjointed and decentralized urban landscapes of late twentieth century North America: vast urbanized regions with varying-sized urban centers, sub-centers and satellites; fragmented and multinodal, with mixed densities and unexpected juxtapositions of form and function.[3] More recently, the term "metroburbia" has emerged in Internet and media usage to capture the way that residential settings in suburban and exurban areas are thoroughly interspersed with office employment and high-end retailing (Figure 3.2).

Today, the traditional form of the metropolis has slipped into history. Contemporary metropolitan America appears to some observers as being characterized by a "splintering urbanism." Michael Dear notes, "It is no longer the center that organizes the urban hinterlands but the hinterlands that determine what remains of the center. The imperatives of fragmentation have become the principal dynamic in contemporary cities. . . . In contemporary urban landscapes, 'city centers' become, in effect, an externality of fragmented urbanism; they are frequently grafted onto the landscape as an afterthought by developers and politicians concerned with identity and tradition. Conventions of 'suburbanization' become redundant in an urban process that bears no relationship to a

FIGURE 3-2. Metroburbia: Fragmented and multinodal mixtures of employment and residential settings, with a fusion of suburban, exurban, and central-city characteristics. This photograph shows part of the Boise, Idaho, metropolitan area. (Heike Mayer)

core-related decentralization."[4] Edward Soja has offered the term "exopolis" in an attempt to capture some of the key dimensions of contemporary urbanization, including the growth of edge cities and the increasing importance of exogenous forces in an age of globalization. Traditional models of metropolitan structure and traditional concepts and labels—city, suburb, metropolis—are fast becoming examples of what sociologist Ulrich Beck calls "zombie categories," concepts that embody nineteenth- to late twentieth-century horizons of experience distilled into *a priori* and analytic categories that still mold our perceptions and sometimes blind us to the significance of contemporary change.[5]

The challenges of characterizing the evolving outcomes of urbanization have in fact prompted a great variety of neologisms, including metroburbia, postsuburbia, exurbia, exopolis, boomburbs, cosmoburbs, nerdistans, technoburbs, generica, satellite sprawl, mallcondoville, and so on. The "New Metropolis," meanwhile, is an encompassing term for the stereotypical urbanized region that has been extended and reshaped to accommodate increasingly complex and extensive patterns of interdependency in polycentric networks of half a dozen or more urban realms and as many as fifty nodal centers of different types and sizes, physically separate but functionally networked and drawing enormous economic strength from a new functional division of labor (Figure 3.3).[6] Bound together through urban freeways, arterial highways, beltways, and interstates, new metropolises are themselves now beginning to coalesce functionally into "megapolitan" regions that dominate national economies.

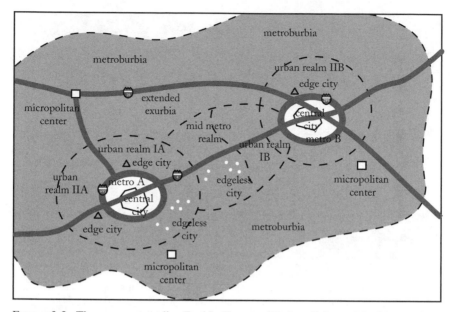

FIGURE 3-3. The new metropolis. (Paul L. Knox and Robert E. Lang, The Metropolitan Institute, Virginia Tech)

The New Metropolis

In many ways, the New Metropolis is a product of the extension and evolution of long-standing elements: demand for low-density single-family housing; access to mortgage finance; economies of scale and scope; fragmentation of land use planning among multiple municipalities; aversion to public transport and the reliance on private automobiles; federally subsidized road building; and exclusionary segregation via deed restrictions. But the New Metropolis is also the product of economic restructuring. Industry, office employment, and retailing had already been significantly decentralized by the 1980s as bigger and faster trucks, larger factories and warehouses, flexible specialization, and just-in-time delivery systems called for a new economic geography that could be accommodated only by larger and less expensive parcels of suburban land. Although heavy industry was usually tied to older, outlying industrial sites with access to rail sidings, most remaining industry had become entirely footloose, able to set up wherever local municipalities could be persuaded to zone land and offer incentives. Corridors of land along major highways quickly became a common setting for footloose industries; industrial parks in campus-like settings became another; while the areas around the junction of major highways and around airports became the preferred location for the burgeoning bioscience and information-based industries, the biggest growth sectors of the modern economy. Meanwhile, along with decentralized population and industry came retailing, service, and office functions: strip malls, big-box discount stores, integrated shopping malls, specialized malls, power centers, fast-food franchises, hotel chains, family restaurants, corporate headquarter complexes, and office parks. By 2000, roughly three out of five jobs in American metropolitan areas were located in, for want of a better name, suburbs.

More important, the *nature* of metropolitan economic development began to change toward the end of the twentieth century with the rapid decline of the old base of manufacturing industries and the onset of a "new economy" based on digital technologies and featuring economic and cultural globalization, the international division of labor, international finance, supranationalism, and cosmopolitanism. Within the United States, capital investment switched increasingly into finance, insurance, and real estate. The structural transformation of the economy resulted in socioeconomic polarization, with increasing numbers of both upper-middle class and impoverished households. The internationalization of economic geography has meanwhile weakened the leverage of both big government and big labor, allowing a fundamental reassertion and intensification of the economic and spatial logic of capital—especially big capital. For some observers, this amounts to nothing less than a second coming of Modernity, a deep structural shift featuring "reflexive modernization" in which the structures and institutions of nineteenth- and twentieth-century modernization are both deconstructed and reconstructed.[7] The ideological shift from egalitarian liberalism to neoliberalism is an important aspect of this. Urban planning has morphed into public-private cooperation; local governments increasingly behave like businesses in their attempts to attract economic development and balance the books; the upper-middle classes have increased their spending power as a result of regressive taxation policies; and a property rights movement has begun to flourish.

The anatomy of the New Metropolis reflects the cumulative impact of these shifts. The country's changing economic geography has meant the decline of older manufacturing districts, especially in the "rustbelt" of the Northeast and Upper Midwest. Downtowns have been reconfigured as they have evolved from hosting the flagship shops and headquarters offices of local firms to accommodating the regional offices of national and international firms. New industries associated with the "new economy" have sought new settings, for the most part well away from congested central city areas where the built environment is ill-suited to their needs. Modern just-in-time production systems and flexible specialization strategies require easily accessible factories; biotechnology firms require specialized new laboratories; while almost every back-office facility and business service requires buildings that are flexible in layout and pre-wired or easily wired for access to digital communications networks.[8] Business and industrial parks are based on single-story structures with designer frontages, loading docks at the rear, and interior spaces that can be used for offices, research and development (R&D) labs, storage, or manufacture, in any ratio. To be competitive, they must also be packaged as "planned corporate environments" with built-in daycare facilities, fitness centers, jogging trails, restaurants and convenience stores, lavish interior décor, and lush exterior landscaping and signage.

The anatomy of the New Metropolis also reflects demographic growth and change. Particularly important in shaping the residential component of the

New Metropolis are "relos," the upper-middle class households that have to relocate as a result of the increasing fluidity and flexibility of corporate location strategies within the new economy. Today's relos are the successors of itinerant white-collar pioneers of the 1960s, like the workers for whom I.B.M. meant "I've Been Moved." They are employees of transnational firms: biomedical salespeople, electronic engineers, information technology managers, accountants, data analysts, plant managers, regional vice-presidents, bankers, manufacturers' representatives, and franchise chain managers,

> the shock troops of companies that continually expand across the country and abroad, [who] move every few years, from St. Louis to Seattle to Singapore, one satellite suburb to another. . . . Relo children do not know a hometown; their parents do not know where their funerals will be. . . . With the spread of global industry's new satellite office parks, the relos churn through towns like Alpharetta; Naperville, Ill., west of Chicago; Plano, Tex., outside Dallas; Leawood, Kan., near Kansas City; Sammamish, Wash., outside Seattle; and Cary, N.C., which is outside Raleigh and, its resident nomads maintain, stands for Containment Area for Relocated Yankees. Converging on these towns, relos have segregated themselves, less by the old barriers of race, religion and national origin than by age, family status, education and, especially, income.[9]

In addition to relos, the residential choices and lifestyle preferences of retiring and empty-nester baby boomers, Gen-Xers, Echo Boomers, and new streams of immigrants all leave their mark on housing markets. With more than 90 percent of metropolitan population growth in America since 1980 having taken place beyond central cities, there is a great deal of socioeconomic diversity across the entire body of the New Metropolis. In conjunction with the complexity and fluidity of economic realignment and restructuring, the uneven evolution of networks of information and communications technologies, and the sheer growth of population, the New Metropolis is characterized by an intense geographical differentiation, and a splintering urbanism featuring metroburbia's new landscapes of innovation, economic development, and cultural transformation, all set amid accelerating sprawl (Figure 3.4).

Metroburbia

The dynamism and restlessness of decentralization, central to the character of the New Metropolis, is frequently depicted in terms of various shock-and-awe statistics. Here are some of them. Between 1980 and 2000, the population of the United States grew by 24 percent, while the extent of urbanized land grew by almost 50 percent; 105 acres of farmland are being withdrawn from agricultural use each hour; about 2.2 million acres—equivalent to thirty-four Orlandos—of wildlife habitat and agricultural land are lost to development each year; every two weeks, Phoenix's built-up perimeter extends by another sixty yards. In short, metropolitan America continues to decentralize, splinter, and sprawl. Among those

FIGURE 3-4. Infill sprawl on the fringes of Boise, Idaho. (Heike Mayer)

metropolitan areas with more than 1,000,000 residents in 2000, forty-three of fifty-three central city cores suffered net domestic migration losses between 2000 and 2005, with a total combined loss of more than 3,250,000. Today, the United States is a metroburban nation with an inner urban periphery and an outer rural fringe. Sprawl itself seems to have splintered into several forms: a recent study of fifty metropolitan areas found four principal forms: leapfrog sprawl; a relatively compact, polycentric sprawl; dispersed sprawl; and a relatively dense and evenly spread form of sprawl.[10] At the local level, the very process of real estate development, uncoordinated and weakly regulated, is predisposed to fragmentation and splintering: "The highway engineers [give] no thought to what would be built at the end of their off-ramps, the subdividers [take] no responsibility for those outside their preferred income niches."[11]

Hopscotch sprawl and proliferating off-ramp subdivisions certainly result in phenomenal rates of growth in some outer metropolitan jurisdictions. Take, for example, Rancho Cucamonga, an hour east of Los Angeles. A sleepy bedroom community of 5,796 in 1970, its population had topped 101,000 by 1990 and by 2005 it had grown to 169,353, all accommodated in a collection of subdivisions with no focal point but, as elsewhere in America, with plenty of familiar landmarks in the form of Home Depot, Lowe's, Wal-Mart, PetSmart, Olive Garden, Circuit City, OfficeMax, Restoration Hardware, Starbucks, Target, Holiday Inn Express, and the rest. Nevertheless, the form of the New Metropolis transcends any simple notion of suburban sprawl and could-be-anywhere strips and malls.

New Metropolitan Form.   The polycentric form of metroburbia is framed around a system of nodes and realms, bound together with ever-expanding four-, six-, and eight-lane highways. It is interspersed with smaller clusters of decentralized employment, studded with micropolitan centers and filled out with booming stand-alone suburbs. Traditional nodal anchors—downtown commercial centers—remain very important, especially as settings for the advanced business services—advertising, banking, insurance, investment management, and logistics services. But in addition there are other nodes within the polycentric form of metroburbia. These vary in character and include:

- "edge cities," decentralized clusters of retailing and office development,[12] often located on an axis with a major airport, sometimes adjacent to a high-speed rail station, always linked to an urban freeway system;
- newer business centers, often developing in a prestigious residential quarter and serving as a setting for newer services such as corporate headquarters, the media, advertising, public relations, and design;
- outermost complexes of back-office and R&D operations, typically near major transport hubs twenty to thirty miles from the main core; and
- specialized subcenters, usually for education, entertainment, and sporting complexes and exhibition and convention centers.[13]

Metropolitan Realms.   Around these nodes are functional realms of various kinds. One example is the "South Coast" realm of the Los Angeles metropolitan region, largely coincident with Orange County. It is clearly part of Greater Los Angeles, but it also maintains a distinct and semi-autonomous identity as "South Coast." Orange County contributes significantly to the region's larger economy but mostly does not compete with Los Angeles. Industries such as automotive design, located at the Irvine Spectrum, a master-planned high tech office park in the center of the county, show this pattern. Several automobile companies chose the Spectrum for access to California trends and regard Orange County as "the next capital of cool."[14] Orange's association with Los Angeles lends a certain chic to this once-sleepy suburban county, while the larger Southern California region gains by additional economic activity.

All realms have their own sub-regional identities. The realms around Los Angeles are so distinctive that South Coast and the Inland Empire (roughly coincident with Riverside and San Bernardino Counties) have their own sub-regional newspapers and airports. On a smaller but emerging scale, a place such as the East Valley of Phoenix (with such major suburbs as Mesa, Tempe, Chandler, and Gilbert) already has its own newspaper and will soon have a separate national airport from Phoenix. Many urban realms also show up in business names, such as South Coast Plaza, Inland Empire National Bank, or the *East Valley Tribune.*

Based on their mix of social characteristics, built densities, and development age, realms can be categorized as follows:

- urban core realms—the original settings for nineteenth- and twentieth-century development, including the region's major principal city and downtown;

- maturing suburban realms—areas of substantial late twentieth-century and early twenty-first-century development that are rapidly filling in;
- favored quarter realms—the most affluent wedge of a metropolitan area, containing upscale communities, luxury shopping centers, and high-end office districts; and
- emerging exurban realms—extended, rapidly growing, lower-density spaces that contain leapfrog development and will not be full extensions of the main metropolitan development for decades to come.[15]

The relationship between these realms plays a role in determining the overall metropolitan dynamic. Favored quarters, such as Southern California's South Coast, are often job rich but have expensive housing. A less affluent maturing suburban realm, such as the Inland Empire, can develop a dependence on a favored quarter. Thus, an important traffic pattern in Southern California is the commute between these two suburban realms. In fact, one of the biggest bottlenecks in the region's freeway system is along a mountain pass (known locally as "The 91") that divides these two realms.

A great deal of employment—especially office employment—is scattered throughout these realms in small office parks and commercial corridors and strips. In fact, only 13 percent of the three billion or so square feet of office space in the thirteen largest U.S. metropolitan areas in 2005 was in edge cities, with 33 percent in downtown settings. Commercial corridors along major intra-metropolitan highways accounted for another 3.8 percent, and urban core realms accounted for an additional 6.4 percent. More than 40 percent of the total office space was scattered in outer suburban and exurban areas, settings that researchers at Virginia Tech's Metropolitan Institute have dubbed "edgeless cities."[16]

Micropolitan Centers.   The residential fabric of the New Metropolis has swollen around this economic framework, filling in the gaps between nodes and corridors of commercial development and sprawling out to peripheral sites in exurban realms. A lot of residential growth has also occurred in and around older, once-distant industrial towns and market towns: not so much edge cities as off-the-edge cities. These are places where middle-class parents still have a chance to buy a spacious four-bedroom house in a town with good, established public schools and wide-open spaces while keeping their metropolitan jobs. Many of them are classified as "micropolitan" by the United States Bureau of the Census. To qualify as micropolitan, an area must have at least one town of 10,000 to 49,999 with proportionally few of its residents commuting outside the area. In 2005, there were 577 micropolitan areas in the continental United States. More than thirty million people, or one in ten Americans, reside in them.

Granbury, Texas, about seventy miles southwest of Dallas, is a good example. Between 1990 and 2000, the population of Granbury and surrounding Hood County increased 40 percent, to 48,000, while median household income jumped from $27,000 to $36,000. "A sparkling lake, affordable housing and the promise of a quieter life have drawn many people to Granbury, refugees both from rural America and suburbia. . . . Telecommuting and Internet mail-ordering can make it easier to be in the fast lane of trade and commerce. Employers find it easier to open a factory or an office park in these towns, which have plenty of room for expansion and low real estate and labor costs, than in traditional metropolitan areas."[17]

Boomburbs. The classic residential component of metroburbia, however, is the "boomburb," defined by Robert Lang and Jennifer LeFurgy as suburban jurisdictions with more than 100,000 residents that have maintained double-digit rates of population growth between 1970 and 2000.[18] According to this definition, there were fifty-four boomburbs in 2000, with an additional eighty-six "baby boomburbs"—places with the same growth qualifications but ranging in size from 50,000 to 100,000 residents. By definition, the most striking attribute of boomburbs is their phenomenal rate of growth. Seven of Lang and LeFurgy's fifty-four boomburbs grew by more than 1,000 percent between 1970 and 2000. Three of these were in the Phoenix region: Gilbert, Arizona, and Peoria, Arizona, recorded over 5,000 percent and 2,000 percent, respectively. But the highest flyer was Coral Springs, Florida: Westinghouse's one-time consumer testing lab logged a remarkable 7,795 percent growth rate, jumping from 1,489 residents in 1970 to 117,549 by 2000.

While boomburbs are found throughout the country, they occur most in the West and Southwest, in a belt of metropolitan areas stretching from Texas to the Pacific, with almost half of them in California alone. In large measure, this is an artifact of administrative boundaries. As Lang and LeFurgy point out, the eastern half of the country has few large incorporated places, outside the traditional central cities, that are able to accommodate the 100,000 threshold. On the other hand, the West has a lot of incorporated places that cover a large territory. The open country of the West also allows developers of large master-planned communities to colonize unincorporated land more easily, and the land and its new residents are frequently added to municipalities, turning what were once small towns into boomburbs. In addition, the public lands in the West that surround big metro areas are often transferred to developers in very large blocks.[19] Lang and LeFurgy suggest that western water districts also play a role in promoting boomburbs. The aridity of the West means that places seeking to grow must organize to access water, and the biggest incorporated places are best positioned to grab a share of water supply. This provides an incentive for fragmented suburbs to join in a large incorporated city.[20] Finally, the local revenue system in many western states relies on municipal-level sales taxes, so the incentive to incorporate land and promote retail development on it has produced what William Fulton calls "sales tax canyons" in places such as Southern California.[21]

The Phoenix metropolitan area includes seven boomburbs, led by Mesa, with almost 400,000 residents. Dallas is also ringed by seven boomburbs, headed by Arlington, whose population grew from 174,299 in 1970 to 333,000

in 2000. Both Mesa and Arlington are already more populous than such traditional hubs as Miami, Minneapolis, and St. Louis. Even such smaller boomburbs
as Chandler, Arizona, and Henderson, Nevada, with 176,581 and 175,381 residents,
respectively, now surpass older mid-sized cities such as Knoxville, Tennessee
(173,890); Providence, Rhode Island (173,618); and Worcester, Massachusetts
(172,648). Los Angeles, with nineteen, has the highest number of boomburbs
and biggest cumulative boomburb population for 2000. But metropolitan Los
Angeles, with over sixteen million people, is so large that its boomburbs account
for just less than one in five residents. At 42.2 percent in 2000, Phoenix has by
far the highest percentage of its metropolitan population living in boomburbs.
The Phoenix area has 1.37 million people living in boomburbs, which slightly
exceeds the number of people living in the city of Phoenix. As in Los Angeles,
these locales account for much of the region's new growth.

Boomburbs are a new and distinctive kind of place, and they are stealthily
eclipsing many traditional cities in terms of economic and demographic vitality.
Plano, Texas, for example, is home to many corporations, including such Fortune
1000 companies as JC Penney and EDS, and contains over eight million square feet
of rental office space—most of it upper-end Class A space. Lang and LeFurgy
note that many boomburbs and baby boomburbs show up in *Money Magazine*'s
list of "Hottest Towns." Eleven of the thirteen top places on the *Money* list in the
West, for example, are boomburbs, starting with Plano, Texas, at number one,
followed by Anaheim, California, in second place, and Scottsdale, Arizona, in third.
Boomburbs typically develop along the interstate beltways that ring large U.S. metropolitan areas, the hubs of the "exit ramp economy,"[22] focused on office parks and
big-box retail strips and centers, all surrounded by subdivisions with large-lot,
single-family homes. While boomburbs possess most "urban" elements—housing,
retailing, entertainment, and offices—they are not typically patterned in a traditional urban form. Boomburbs almost always lack, for example, a dense or identifiable business core. Boomburbs can thus be seen as distinct from traditional urban
centers not so much in their function as in their low density and loosely configured
spatial structure. "Boomburbs are urban in fact, but not in feel."[23]

Exurbs. A more general term, encompassing parts of many boomburbs
but extending to still more peripheral settings, is "exurb." "The exurbs," argues
Anthony Flint, "are without question the new sociological phenomenon of our
time. Americans are moving there in droves, to Pasco County, Florida, and all
around Boise, Idaho, parking their cars in short driveways and never venturing into a city, ever. They are leading a self-contained life, with work at an
office campus and big-box stores along a commercial strip all requiring lots of
motoring."[24] The term "exurb" was coined in the 1950s in *The Exurbanites* by
A. C. Spectorsky, a social historian, to describe semi-rural areas far outside cities
where wealthy people had country estates.[25] Today's exurbs are populated by the
world's first mass upper-middle class, living in two-, five-, and ten-acre "ranchettes"
and in subdivisions of large homes, all set amid landscapes that still contain

extensive tracts of rural farmland and pockets of smaller homes owned by working-class families along the highways and on one- or two-acre lots tucked away on back roads.

The expansion of suburbs and exurbs since 1960 has been dramatic, but defining and measuring exurbia with any precision is challenging. Remote sensing from satellites shows exurbia to be extensive.[26] Detailed maps produced by David Theobald define exurbia in terms of housing density, using block-level census data. Theobald classifies "urban" (that is, urban and suburban) densities as fewer than 1.7 acres per housing unit, and "exurban" as 1.7 to 40 acres per unit (see Figure 3.5).[27] A more recent Brookings Institution study has defined exurbs in terms of commuting (at least 20 percent of workers commuting to an urbanized area within a large metropolitan area) and growth rates (population growth between 1990 and 2000 that exceeded the average for its related metropolitan area) as well as housing density (2.6 acres or more per unit). According to this definition, approximately 10.8 million people lived in the exurbs of large

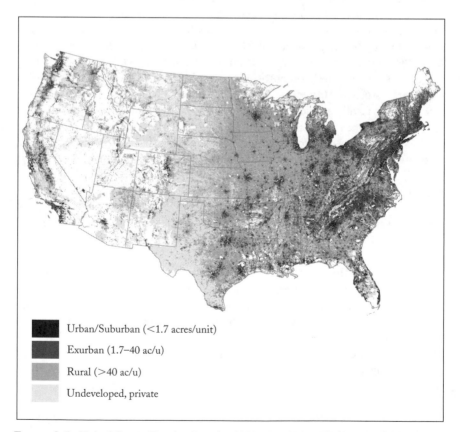

Urban/Suburban (<1.7 acres/unit)

Exurban (1.7–40 ac/u)

Rural (>40 ac/u)

Undeveloped, private

FIGURE 3-5. United States Housing Density 2000. (David M. Theobald, Colorado State University)

metropolitan areas in 2000—roughly 6 percent of their population. These exurban areas grew more than twice as fast as their respective metropolitan areas overall, while residents of the "average" exurb were disproportionately white, middle income, homeowners, and commuters.[28] In Chicago (Figure 3.6), the study found a considerable number of exurban communities in former farmlands throughout the metro area, and in some instances, beyond. The region's exurbs extend into southern Wisconsin and western Indiana; more than half the residents of Jasper County, Indiana, live in fast-growing, low-density areas with significant commuting ties to the Chicago area's urban core. In the more markedly polycentric Los Angeles metropolitan region (Figure 3.7), nodes of exurban development are evident in southern Orange County around Laguna Niguel; in northern

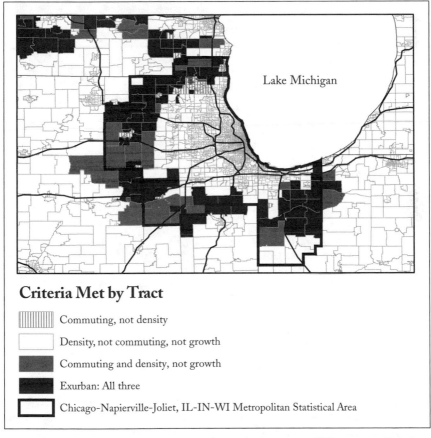

Criteria Met by Tract

||||||||| Commuting, not density

Density, not commuting, not growth

Commuting and density, not growth

Exurban: All three

Chicago-Napierville-Joliet, IL-IN-WI Metropolitan Statistical Area

FIGURE 3-6. Exurban character of census tracts in the greater Chicago area. Based on "Commuting Ties, Population Growth, and Housing Density," in Alan Berube, Audrey Singer, Jill H. Wilson, and William H. Frey, *Finding Exurbia: America's Fast-Growing Communities at the Metropolitan Fringe* (Washington, D.C.: Brookings Institution, Living Cities Census Series, October 2006), Map 2, p. 15. www.brookings.edu/metro.

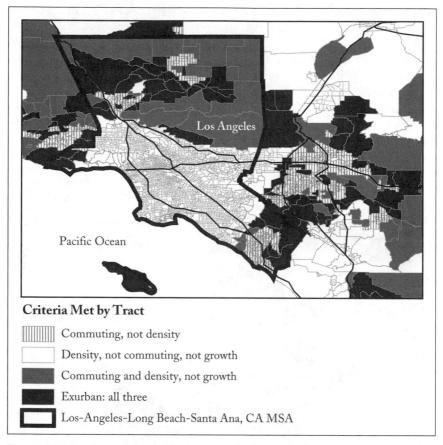

FIGURE 3-7. Exurban character of census tracts in the greater Los Angeles area. Based on "Commuting Ties, Population Growth, and Housing Density," in Alan Berube, Audrey Singer, Jill H. Wilson, and William H. Frey, *Finding Exurbia: America's Fast-Growing Communities at the Metropolitan Fringe* (Washington, D.C.: Brookings Institution, Living Cities Census Series, October 2006), Map 3, p. 16. www.brookings.edu/metro.

Los Angeles County outside of Lancaster and Palmdale; in the outskirts of Simi Valley and other areas surrounding Thousand Oaks; and in inner Riverside County near Corona and Ontario. Measured as a proportion of metropolitan population, Washington's exurbs (Figure 3.8) are among the largest. They are also far-flung, reaching into Spotsylvania County, Virginia, where they begin to bleed into Richmond's northern exurbs; west into Warren County, Virginia, at the base of the Shenandoah Mountains; north into Frederick County, Maryland; and to the southeast in Calvert, Charles, and St. Mary's counties, Maryland. Baltimore's exurbs meet Washington's in fast-growing western Howard County, Maryland.

Exurbia is somewhat constrained in the West and Southwest, where water is in short supply and public agencies and utility companies cannot afford to

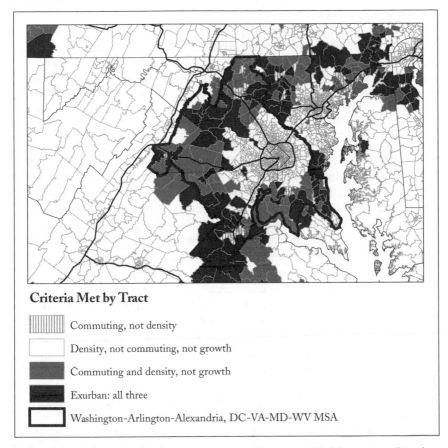

Criteria Met by Tract

Commuting, not density

Density, not commuting, not growth

Commuting and density, not growth

Exurban: all three

Washington-Arlington-Alexandria, DC-VA-MD-WV MSA

FIGURE 3-8. Exurban character of census tracts in the greater Washington area. Based on "Commuting Ties, Population Growth, and Housing Density," in Alan Berube, Audrey Singer, Jill H. Wilson, and William H. Frey, *Finding Exurbia: America's Fast-Growing Communities at the Metropolitan Fringe* (Washington, D.C.: Brookings Institution, Living Cities Census Series, October 2006), Map 4, p. 17. www.brookings.edu/metro.

overextend their systems. The most striking examples of exurban growth are therefore in the East and Northeast. Caroline County and King George County, for example, some sixty-five miles from downtown Washington in the Piedmont region of Virginia, are now among the nation's fastest-growing counties. Until recently they were mostly rural, sparsely populated, with landscapes of farms and pine forests, cement and lumber plants, travel plazas, biker bars, truck washes, cheap townhouses, and Baptist churches. A few long-distance commuters to Richmond populated a semblance of suburbia. Now, the landscapes of both counties are being rapidly rewritten with new ranchettes and upscale subdivisions of faux French Country and neo-Georgian homes occupied by federal and high-tech workers from the national capital region.

It is the fate of exurbs eventually to become more generically suburban. This is what has already happened, for example, to Fairfax County, Virginia, just outside Washington's beltway and one of the nation's fastest-growing counties in the 1980s and early 1990s. Once—in the 1970s—a classic exurban fringe, Fairfax County is now effectively built out, with new development strictly controlled. In the 1980s, Fairfax County took out double-page advertisements in the *Wall Street Journal* touting the county's virtues as a locale for corporate headquarters— a costly and, at the time, virtually unheard-of campaign of self-promotion for a suburban locality. The locus of the subsequent boom in Fairfax was the Dulles corridor, along the highways that run from Tysons Corner—the Washington Metropolitan area's largest edge city—to Dulles International Airport. Very quickly, thousands of new employees working in the Dulles corridor were priced out of living in Fairfax County. Then, alarmed at the pace of residential growth, Fairfax leaders downzoned extensive swaths of the residential-zoned land in the county to no more than two homes per acre, citing reasons of environmental, agricultural, and neighborhood preservation. The subtext was that while new jobs meant revenue for the county, new houses meant intensified demand for services and schools, which drain the county exchequer and drive up local taxes.

For workers in and around the Dulles corridor, Fairfax's restrictions helped make neighboring Loudoun County a logical place to look for a home. Immediately to the northwest of Fairfax, Loudon county is still one of the ten fastest-growing counties in the nation but it too is in the process of filling up: the new battleground for no-growth and slow-growth activists. Loudon had briefly been exurban territory, but during the 1990s the number of Fairfax workers living in Loudoun doubled, helping to increase the county's population from 86,000 to 170,000. This influx into Loudoun, however, set off a political reaction. Led by a citizens group that called itself Voters to Stop Sprawl, voters installed a slow-growth set of leaders in the 1999 elections. A fiscal impact analysis suggested that the county could save $103 million annually by 2020 by slowing residential growth. Within two years, Loudoun's county board had reduced the number of houses to be permitted within its jurisdiction from about 187,000 to about 100,000 while effectively placing the western two-thirds of the county off limits to conventional subdivisions, with developers generally limited to no more than one home per ten acres.[29] Cutting the number of homes reduced the demand for schools and other services that can drain tax dollars. The county board wanted businesses to come to Loudoun—"boardrooms, not bedrooms"—because they typically generate more in taxes than they consume in services.

The locus of Washington's western exurban development has now shifted beyond Loudoun County and into the northeastern tip of West Virginia. In Charles Town, West Virginia (population 2,907 in 2000), for example, the new 3,200-home Huntfield community stands more than twenty-five miles from the nearest job center, in Leesburg, and twice as far from offices in Reston, Tysons Corner, and Washington. Yet, according to statistics provided by the developer, only one of the first 100 homebuyers will work in West Virginia. Of the others, seventy-two will work in suburban Virginia, thirteen in Maryland, and five in Washington.[30] The rest identified themselves as self-employed or retired. The daily commute for most of these residents takes a minimum of an hour and a quarter; the perceived trade-off is the saving on Huntfield's new four-bedroom homes that sell for about $270,000—at least $150,000 less than the cost of a similar house closer to Washington. These savings may or may not turn out to be real over the longer term. An analysis of twenty-eight metropolitan areas by the Center for Housing Policy found that the costs of one-way commutes of as little as twelve to fifteen miles cancel any savings on lower-priced outer-suburban homes.[31]

There are parallels to these examples everywhere. In the southern reaches of the Washington metropolitan region, exurbia has moved on from Prince William County and Stafford County to the exurban territory of Caroline County and King George County described above. In the Atlanta region, Gwinnett County, formerly classic exurban territory, is now a sea of subdivisions and master-planned communities with a population approaching half a million. Seen from the air, it appears as a vast labyrinth of cul-de-sacs—an estimated 9,000 of which are laced throughout the county.[32]

The New Scale: Megapolitan

Perhaps the most distinctive attribute of the New Metropolis, its signature feature, is its sheer scale. Bound together through urban freeways, arterial highways, beltways, and interstate highways, the prototypical New Metropolis is rapidly emerging as part of a megapolitan region.[33] Megapolitan regions are integrated networks of metropolitan areas. According to Lang and Nelson, the United States has twenty megapolitan regions (see Figure 3.9).[34] As of 2006, six in ten Americans, or more than 180 million people, lived in these megapolitan areas, which between them covered only a tenth of the nation's land area. "Megas account for nearly three quarters of US gross domestic product. Within their space lies the nation's leading office markets and its high tech heartlands including Boston's Route 128, the Bay Area's Silicon Valley, Northern Virginia's Dulles Toll Road, and Austin's Silicon Prairie."[35] By 2040, megapolitan regions are projected to gain over eighty-five million residents, or about three-quarters of national growth.[36] Much of this development will fill in the remaining gaps between metropolitan areas, consolidating the links among principal cities and micropolitan areas within megapolitan regions.

Interstate highways are key structural elements in megapolitan development. Interstate 95, for example, plays a major role in structuring the New England Megalopolis, the Mid-Atlantic Megalopolis, the Chesapeake Megalopolis, and, further south, Florida's Gold Coast. The West's equivalent of Interstate 95 is Interstate 5, which runs through three separate megapolitan regions. Interstate 10 also structures—and links—several megapolitan regions: Southern California, the Sun Corridor, the Texas Corridor, and the Texas Gulf. Interstate 85 forms the backbone of the Carolina Piedmont and Southern Piedmont regions, running from

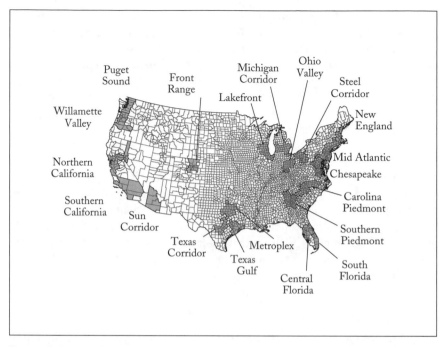

Figure 3-9. America's megapolitan regions. (Robert E. Lang and Arthur C. Nelson, The Metropolitan Institute, Virginia Tech)

Raleigh, North Carolina, southwest to Atlanta, Georgia. Megapolitan regions vary in spatial form and scale. Some exhibit a corridor (or linear) form, while others spread out into vast urban galaxies. Arizona's Sun Corridor (see Figure 3.10), for example, exhibits a nascent corridor structure and covers a fairly modest area, while the Carolina Piedmont region (Figure 3.11) has a "galactic" form.

Cosmoburbs

The scale and extent of the New Metropolis inevitably encompasses a great diversity of populations. No longer is suburbia synonymous with affluent and upwardly mobile white nuclear families. Nor should we continue to retain the image of suburbs as growing at the expense of central-city decline. As metropolitan areas have grown, their suburbs have changed, becoming quasi-urbanized. Moreover, while suburban population as a whole has been growing, it has been highly uneven from suburb to suburb, metropolis to metropolis. A comprehensive analysis of 2,586 suburban municipalities by William Lucy and David Phillips found that 700 of them—27 percent—actually declined in population between 1990 and 2000, with 124 of them—almost 5 percent of the total—declining in population faster than their central cities. Lucy and Phillips also showed that, contrary to the popular image of suburbia as solidly white, upper-middle-class, and family-oriented—a stereotype that is still marketed vigorously

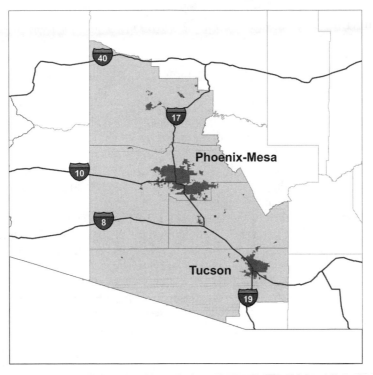

FIGURE 3-10. Arizona's Sun Corridor. (Robert E. Lang, The Metropolitan Institute, Virginia Tech)

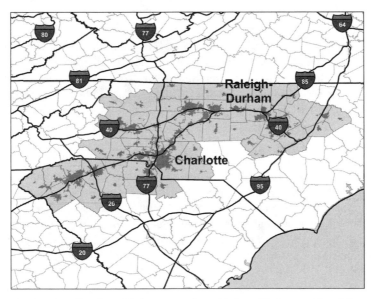

FIGURE 3-11. The Carolina Piedmont region. (Robert E. Lang, The Metropolitan Institute, Virginia Tech)

to the public by developers—the suburbs contain significant and increasing numbers of ethnic minorities, including many immigrants, along with significant concentrations of poverty and a growing share of the nation's single-person households and seniors.[37]

According to an analysis of census data by William Frey and Alan Berube, minority groups constituted the largest demographic shift from cities to suburbs in the 1990s. In 65 of the largest 102 metropolitan areas, "minority flight" equaled or outpaced "white flight" to the suburbs.[38] Hispanic and Asian gains in the suburbs, meanwhile, were largely due to new arrivals who have jumped past central cities. Today, half of all immigrants move directly to suburbs. Asian immigrants, more than any other group, select suburban lifestyles. There are, of course, many affluent suburban communities that are almost entirely white, but there are also affluent African-American suburbs, Chinese-American suburbs, and Korean-American suburbs, as well as ethnically and demographically diverse suburbs. Robert Lang and Jennifer LeFurgy have coined the term "cosmoburbs" to describe wealthy suburbs that are also diverse. Leading examples around the nation include Irvine, California; Carrolton, Texas; and Bellevue, Washington.[39] Lang, together with Ed Blakely, takes issue with the conventional stereotype of Orange County, California ("the OC"), as portrayed in the media as a locus of affluent white suburbia. They point out that Orange County is not even majority white and argue that Anaheim, in the heart of the county, has grown so diverse that it qualifies as a "New Brooklyn," a large, formerly white suburb with a high percentage of foreign-born residents—exactly the same, in fact, as the old Brooklyn (38 percent). "Anaheim's Levitt-type subdivisions—built around the time that Disneyland opened in the 1950s—exemplified the period's American Dream of a home in the suburbs, which was then limited by covenant to whites. Today, those neighborhoods are home to most of the city's new immigrants. The post-war Anaheim suburban landscape now comprises some of America's most dynamic and offbeat places."[40] Neighboring Santa Ana, the county seat of Orange County, has a population that is over half foreign-born (53 percent). Moreover, the mostly affluent and master-planned Orange County community of Irvine, home of a University of California campus, is just under one-third foreign-born (32 percent).

While immigration and residential mobility have been changing the face of metroburbia, changes in the economy have led to a structural shift in income distribution. The "new economy" is an "hourglass economy," increasingly polarized in terms of income distribution. In 1985 there were just thirteen U.S. billionaires. Now there are more than a thousand. The super-rich are getting richer faster than ever, assisted by fluid and speculative markets and corporate tax breaks granted by the George W. Bush administration. In 2005 the top 1 percent of earners in the United States gained 14 percent in income in real terms, while the rest of the country gained less than 1 percent. That same year, 227,000 people became millionaires. The phenomenal incomes and attention-getting extravagance of America's super-rich have been described by Robert Frank in his book

Richitsan.[41] Meanwhile, the number of persons in the United States living below the poverty line—thirty-six million—is at a thirty-three-year high.

The middle-middle class has declined relative to both the upper-middle class and the poor, and this has inevitably been reflected in the suburbs' fiscal and social conditions. An analysis of the twenty-five largest metropolitan areas in the United States classified 55 percent of the suburban population as living in "at-risk" suburbs, 36 percent in "bedroom-developing" suburbs, and 9 percent in "affluent job centers."[42] Only the affluent job centers have adequate infrastructures and public sector resources. The U.S. Department of Housing and Urban Development, in its 1999 report on *The State of the Cities,* warned that conditions commonly associated with troubled inner cities had been discovered in nearly 400 suburbs, including poverty rates of more than 20 percent.[43] Lucy and Phillips found that the number of low- and moderate-income suburbs (below 80 percent relative median family income) increased by a factor of four between 1960 and 1990, while concentrated poverty in suburban neighborhoods nearly doubled between 1980 and 2000. Approximately half of the 2,586 suburban municipalities in their analysis had relative incomes below those of their respective metropolitan-wide median incomes in 2000, and about half had declined in relative income between 1990 and 2000. This included some suburbs as much as forty miles from city centers as well as inner-tier suburbs. At the other end of the scale, among the 50 percent or so of suburbs where relative incomes had increased, increasing affluence was particularly pronounced in rapidly growing peripheral suburbs.[44]

The Secession of the Successful

Amid this reality of diversity, polarization, and decline, the most affluent and aggressively aspiring of America's households must struggle to establish their place in metroburbia. Rising incomes at the top of the hourglass, flattening prices, and easily available credit have given upper-middle class Americans access to such a wide array of high-end goods that traditional markers of status have lost much of their meaning. The problem is acute for the higher-earning professional and semi-professional classes, credentialed workers and relos, and especially so for "symbolic analysts"—such as management consultants, lawyers, software and design engineers, research scientists, certain medical specialists, therapists, corporate executives, financial advisers, strategic planners, advertising executives, television and movie producers, and various kinds of consultants—whose jobs consist of analyzing and manipulating symbols, words, numbers, or visual images. As Barbara Ehrenreich pointed out, because expertise—the currency of these symbolic analysts—cannot be passed on or inherited by the next generation like wealth or property, this class fraction lives with a constant "fear of falling" from its privileged place in society. Parents are all too aware that there is little they can do to guarantee their children a comfortable life beyond encouraging a fierce appetite for success. Little wonder that we find über-moms—"blonde ponytail, dark sunglasses, yoga pants, hurtling toward junior's violin lesson in her black SUV; Whole Foods chai in her cupholder" and "with their shining yoga faces, Mozart in the womb, natural birth, . . . getting on extended waiting lists for super pre-schools, piano lessons at two."[45] In Los Angeles, some private pre-schools have Admissions Coordinators and Academic Performance Indexes that predict children's future academic success; "tuition" for kindergarteners can run $14,000 to $26,000 a year.

The collective status anxiety of this class fraction has expressed itself in numerous ways, with the result, says Ehrenreich, that it "plays an overweening role in defining 'America': its moods, political direction, and moral tone."[46] Their class-consciousness, Ehrenreich argues, has not only estranged them from "ordinary" Americans but has also brought with it hedonism, self-indulgence, and a pervasive and deep-seated anxiety. Through professionalization and credentialism, the symbolic analysts of the upper-middle class had sought to establish their distinction in an increasingly uncertain world. But they quickly learned that the barriers erected to exclude other classes also stood in the way of their own children, leaving them in a fragile position, their distinction and identity threatened. Traditional identifiers that had served core professions such as doctors, lawyers, architects, and teachers well in the past—liberalism, estheticism, moral integrity, counterculturalism, and intellectualism—had been effectively closed off by the 1980s as neoconservatives successfully branded them as conspiratorial and anti-American. As a result, argues Ehrenreich, the rapidly expanding class fraction of new professionals and symbolic analysts turned to what she calls the "yuppie strategy": the superficial emulation of the truly wealthy through hedonistic and competitive consumption. Self-absorbed, unable to look beyond their own neuroses, devoting ever-increasing levels of energy to occupational achievement and self-satisfaction, they have little left over for civic leadership or social concerns. The "fear of falling" identified by Ehrenreich has a lot to do with the attenuation of the social compact that was the hallmark of industrial society and the emergence of a post-industrial "risk society" in which class conflicts and political cleavages are increasingly defined in terms of exposure to hazards and risks.[47]

This translates directly into the social ecology of metroburbia. Richard Sennett contends that "the way cities look reflects a great, unreckoned fear of exposure. 'Exposure' more connotes the likelihood of being hurt than of being stimulated. . . . What is characteristic of our city-building is to wall off the differences between people, assuming that these differences are more likely to be mutually threatening than mutually stimulating. What we make in the urban realm are therefore bland, neutralizing spaces, spaces that remove the threat of social contact."[48] Robert Reich takes this further, pointing to the "seccession of the successful"—the top quintile of the socioeconomic ladder who take home more money than the other four-fifths put together. Their secession from the rest of the New Metropolis, argues Reich, takes several forms:

> In many cities and towns, the wealthy have in effect withdrawn their dollars from the support of public spaces and institutions shared by all

and dedicated the savings to their own private services. As public parks and playgrounds deteriorate, there is a proliferation of private health clubs, golf clubs, tennis clubs, skating clubs and every other type of recreational association in which costs are shared among members. Condominiums and the omnipresent residential communities dun their members to undertake work that financially strapped local governments can no longer afford to do well—maintaining roads, mending sidewalks, pruning trees, repairing street lights, cleaning swimming pools, paying for lifeguards and, notably, hiring security guards to protect life and property.[49]

In some cases, this impulse can lead to attempts to secede formally and politically from existing municipalities, as in the case of the residents of affluent portions of the San Fernando Valley in the Los Angeles metropolitan area in 2002, organized as Valley Voters Organized Toward Empowerment (Valley VOTE). Municipal fragmentation in the suburbs is by no means new, of course: it was a tenet of public choice theorists in economics that smaller units of government deliver services more efficiently and are more responsive to constituent needs. In this view, local governments are essentially service providers that compete for residents by offering various packages of city services. When individuals are no longer satisfied with a city's service offerings, residents can "vote with their feet" for a more amenable jurisdiction.[50] The resulting competition between localities, it is alleged, forces cities to cut waste and improve efficiency. But examples like Valley VOTE are about much more than the fiscal mercantilism of collective attempts to maximize bundles of public goods while minimizing tax burdens: "In other words, it is not that wealthy suburban enclaves just want to exclude the poor and minorities, but they want to exclude anyone."[51]

Secession into Lifestyle Enclaves

More frequently, the impulse to secession leads households to gated communities and homeowner associations, settings where the top quintile can pursue the American Dream Extreme. In a climate of heightened perceived risk, and in a society where orientation, direction, and purpose are increasingly attenuated and fractionalized, "forting up" in walled and gated communities provides a substitute for the ontological security[52] of the old order. In these settings, the new professional classes, relos, and symbolic analysts must live without the traditional community props of the upper-middle classes: pedigree and family ties; seats on the vestry and the hospital board; inherited wealth and property; and the rituals, like charity balls, silent auctions, and hunting weekends. They are left with their emblematic icons of consumption: Lexus and Mercedes automobiles, Etro skirts and Ermenegildo Zegna shirts, Manolo Blahnik shoes, Kate Spade and Judith Leiber handbags, Pamela Kline soft furnishings, Dransfield and Ross bed linens, Callaway golf clubs, and, most conspicuously, their houses. Houses, for this class fraction, have come to represent a particularly virulent form of what Dejan Sudjic has called the "edifice complex," the use of built form by the rich and powerful to impress and intimidate.[53]

In addition to their ostentation through sheer scale and superfluity—panic rooms, multimedia theaters, and restaurant-sized kitchens—upscale residences also serve as proscenia for the aspirational lifestyles of their inhabitants: a topic that is explored in detail in chapter 7. For the moment, it is sufficient to note that the splintering urbanism that is characteristic of metroburbia has generated a mosaic ecology of lifestyles, all of which have been squarely caught in the crosshairs of market researchers. Euro RSCG, the world's fifth-largest advertising agency and popularizer of the term "metrosexual," has broken down upscale American mothers into the following "hot" advertising categories: Domestic Divas (who employ nannies to raise flawless kids and housekeepers to keep their homes gorgeous), Boomerang Moms (who worked when their kids were small and left work when kids were teens), Yummy Mummies (staying fabulous at the gym), and Mini-Me Moms (whose kids are fashion accessories).[54] Among the sixty-six neighborhood lifestyle types identified by the Nielsen corporation's Claritas PRIZM®NE marketing system[55] are at least twenty-five that are associated with the suburbs, exurbs, and satellite centers of the New Metropolis. These are described in chapter 7.

Making Room at the Top: The Leading Edge of Suburbia

The one-fifth of American households at the top of the hourglass economy had incomes in excess of $91,000 in 2005. Within metropolitan regions, the incomes of the most affluent households are higher: in the Washington metropolitan region, for example, median household incomes in the top quintile of census tracts in 2000 ranged between $84,973 and $232,608; in Chicago they ranged between $68,810 and $350,000 (see Table 3.1). Households at the top of the hourglass economy have been racing ahead of the other four-fifths in the past decade or two: the number of households with incomes of more than $100,000 grew 29 percent more than the growth in the number of U.S. households between 1986 and 1995, and 26 percent more than the growth in the number of U.S. households between 1996 and 2005. This was at a time when mortgage interest rates hit their lowest levels in more than four decades. The result was a tremendous increase in new home building at the top end of the market. After the

TABLE 3.1.   *Median Household Income of the Top Quintile of Census Tracts, Selected Metropolitan Regions, 2000*

Metro region	Range	Median for top-quintile tracts
Chicago	$68,810–$350,000	$74,527
Dallas	$68,515–$242,948	$86,601
Denver	$75,843–$203,060	$86,839
San Bernardino–Riverside	$57,072–$95,832	$69,202
Washington, D.C.	$84,973–$232,608	$103,603

stock-market bubble of the late 1990s burst, the trend intensified still further as more people decided that the best bet for their investments was real estate. Inevitably, house prices soared; but mortgage financiers found increasingly novel ways to package loans to help people afford houses that otherwise would be beyond their reach: interest-only mortgages, graduated-payment mortgages, growing-equity mortgages, jumbo and super-jumbo mortgages, shared-appreciation mortgages, step-rate mortgages as well as sub-prime mortgages that supercharged the market and contributed to the housing market bubble of the first half of the 2000s.

Some of the building at the top end of the market was in the form of in-town developments: luxury downtown apartment blocks, monster homes replacing tear-downs in inner-suburbs, condos in waterfront and riverside redevelopments, and so on. The 1990s and early 2000s housing boom coincided with a significant "back-to-the-city" movement, especially among "empty-nester" baby boomers.[56] But most of the residential construction took place in the interstices and sub-urban fringes of the New Metropolis, vaulting neighborhoods from relative anonymity and into the top socioeconomic bracket. Figures 3.12 through 3.16 show census tracts in the Chicago, Dallas, Denver, Riverside–San Bernardino, and

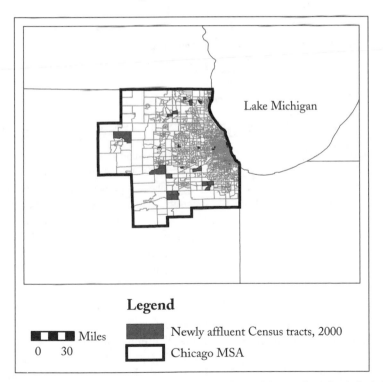

FIGURE 3-12. The secession of the successful: Chicago Metropolitan Statistical Area. Census tracts that were in the top quintile in the region in terms of median household incomes in 2000 but that had never previously (in the 1970, 1980, or 1990 census returns) been so.

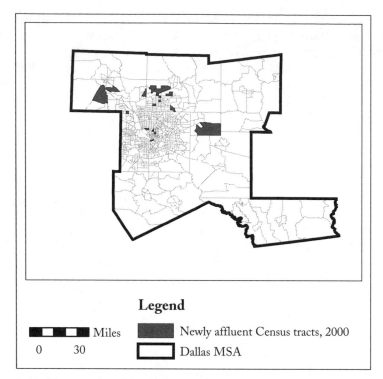

FIGURE 3-13. The secession of the successful: Dallas Metropolitan Statistical Area. Census tracts that were in the top quintile in the region in terms of median household incomes in 2000 but that had never previously (in the 1970, 1980, or 1990 census returns) been so.

Washington, D.C., Metropolitan Statistical Areas (MSAs) that were in the top quintile in terms of median household incomes in 2000 but had never previously (in the 1970, 1980, or 1990 census returns) been so. These represent the leading edges of suburbia, the places that have been developed through the secession of the successful.

These maps show that newly affluent neighborhoods are not found in any single sector or zone in any of these metropolitan areas—as might have been expected from traditional models of urban change. Rather, the pattern supports the notion of splintering urbanism. In general terms, three kinds of localities show up on these maps:

1. Central city census tracts where redevelopment has vaulted part of an older neighborhood into the top quintile;
2. Suburban and exurban census tracts where upscale master-planned communities have recently been developed on greenfield sites; and
3. Peripheral census tracts that have recently become part of exurbia, with scattered, low-density development of custom and semi-custom homes.

In the Chicago Metropolitan Statistical Area (Figure 3.12), almost all of the leading edge of affluent suburban development is of the second type: new

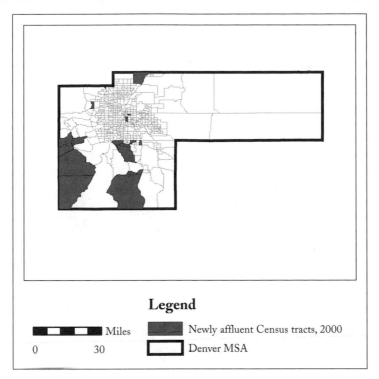

Legend

Miles

0   30

Newly affluent Census tracts, 2000

Denver MSA

FIGURE 3-14. The secession of the successful: Denver Metropolitan Statistical Area. Census tracts that were in the top quintile in the region in terms of median household incomes in 2000 but that had never previously (in the 1970, 1980, or 1990 census returns) been so.

master-planned communities, many of them with private golf courses. These include extensive tracts of development near Oswego and between Plainfield and Romeoville in the southwest of the metro region; and near the Fox River in the north of the region, around Foxford Hills, Tower Lakes, and Fox River Valley Gardens. In the far western reaches of the region, a new patch of exurbia, with scattered single-family homes, sits along Route 72 in De Kalb County, Illinois, between Kirkland and Genoa. Similarly, in the Dallas Metropolitan Statistical Area (Figure 3.13) there is one new patch of exurbia (east of Lake Hubbard, on the eastern fringe of the metro region, between Heath and McLendon-Chisholm) but most of the leading edge of affluent suburban development is in the form of master-planned communities. These are mostly to the north and northwest of the central city and include communities along the northwestern shores of Grapevine Lake and a series of communities along Highway 121 such as The Colony and the Plantation Resort Golf Club community.

In the Denver-Boulder-Greeley Metropolitan Statistical Area (Figure 3.14), a mixture of exurban development and master-planned developments follows Interstate 25 south from Denver, while to the north of the city there is an outlier

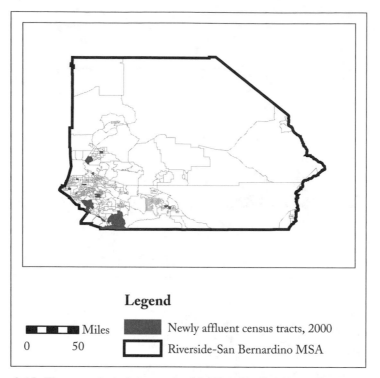

Legend

▮▮▮▮ Miles

0   50

▮ Newly affluent census tracts, 2000

☐ Riverside-San Bernardino MSA

FIGURE 3-15. The secession of the successful: Riverside–San Bernadino Metropolitan Statistical Area. Census tracts that were in the top quintile in the region in terms of median household incomes in 2000 but that had never previously (in the 1970, 1980, or 1990 census returns) been so.

of master-planned development around Todd Creek, just to the west of Brighton. In California, virtually all the areas of fast-paced economic growth in the state are located in the "Inland Empire" of Riverside and San Bernardino counties (Figure 3.15). Vast tracts of the Riverside–San Bernardino Metropolitan Statistical Area remain undeveloped but, increasingly, western Riverside and San Bernardino Counties are featuring the type of upscale houses, stores, and entertainment long found in Los Angeles and San Diego. Tall office buildings are sprouting, along with more $1-million-plus homes, and the region's median income now surpasses that of Los Angeles County. The leading edge of affluent suburban development, as reflected in Figure 3.15, is scattered principally along the Chino Hills and the hillsides that border the Corona Freeway. Much of this is exurban settlement, as are the outliers to the north of San Bernardino, near Interstate 15. About fifty-five miles south-southeast of San Bernardino, along Interstate 10 toward La Quinta, are small outliers dominated by master-planned communities, including Palm Desert Resort Country Club, Bermuda Dunes Country Club, Lakes Country Club, and Mountain Vista Golf Club communities.

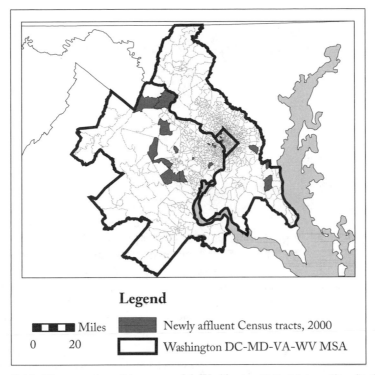

Legend

▮▮▮ Miles   ▮ Newly affluent Census tracts, 2000

0   20   ▭ Washington DC-MD-VA-WV MSA

FIGURE 3-16. The secession of the successful: Washington, D.C., Metropolitan Statistical Area. Census tracts that were in the top quintile in the region in terms of median household incomes in 2000 but that had never previously (in the 1970, 1980, or 1990 census returns) been so.

The Washington, D.C., metro region experienced its first real private-sector boom in the mid- to late 1990s, with technology companies including America Online and a host of telecommunications start-up businesses popping up, particularly in northern Virginia, fueling the exurban and suburban development in Fairfax and Loudoun described above.[57] On the Maryland side, the development of a medical and biotechnology corridor between Rockville and Gaithersburg had a similar effect. The leading edge of affluent suburban development in the Washington D.C. Metropolitan Statistical Area is located principally in patches of master-planned developments in Prince William County, Virginia (between Gainesville and Bristow, and along the new Prince William Parkway, between Manassas and Woodbridge), Loudoun County, Virginia (along the Dulles Greenway, west of Dulles International Airport), and Montgomery County, Maryland (along Interstate 20 north of Gaithersburg, between Germantown and Clarksburg).

These new frontiers of the New Metropolis are the focus of much of the rest of this book. We begin with an examination of the influence of key "place entrepreneurs" in the home-building industry and a review of the "sprawl machine," the influence of big development companies on recent residential development trends.

CHAPTER 4

Developers' Utopias

Just as metropolitan form is constantly evolving, so too is the development industry that produces it. Property development involves many different actors and institutions, and the nature of their involvement can vary from one project to another. Landowners, subdividers, investors, financial analysts, commercial banks, title insurance and trust companies, mortgage companies, lawyers, federal, state, and local agencies, transportation and utility companies, architects, landscape architects, planners, civil engineers, appraisers, general building contractors, subcontractors, realtors, market researchers, special-interest groups, political leaders, and consultant media, marketing, and imagineering specialists are all involved in property development, but the principal actors, the key "place entrepreneurs," are developers. Some developers are involved principally with land, some with buildings, and many with both. Land developers typically acquire raw or unimproved land and improve it with earth grading, roads, utility connections, deed restrictions, and entitlements.[1] Building or project developers acquire improved land or redevelopable property and construct new buildings. Metroburbia is largely a product of big developers whose firms are not only involved with land and building but also with many aspects of finance, insurance, marketing, and local politics.

Historically, real estate development in American cities has been a predominantly local affair, organized on a project by project basis by real estate promoters, financiers, or investors and implemented under contract by small, local construction firms. In 1929, for example, only 2.5 percent of the high-volume builders in the United States were "operative builders" (also known as merchant or speculative builders) who constructed buildings for their own account rather than under contract, in order to sell the completed building(s) for profit on a fee-simple basis.[2] "These were the cream of their trade, occasionally venturing to erect office or apartment buildings, but generally concentrating on speculative homebuilding."[3] The stabilization of the mortgage market and the setting of minimum standards for the housing financed by the Federal Housing Administration in the 1930s allowed more and more operative builders to become what Marc Weiss calls "community builders"—developers who subdivide and improve raw land and design, engineer, finance, construct, and sell buildings on the extensive sites that they have prepared.[4] These community builders were the precursors of the developer-builder companies that now dominate the design and construction of the fabric of metroburbia. It was the community builders of the 1930s and 1940s who pioneered deed restrictions mandating uniform building lines, front and side yards, standards for lot coverage and building size, and minimum construction standards, as well as innovations in landscaping, street layout, and planned provision for retail and office buildings, parks and recreation facilities, churches and schools. After World War II, it was the community builders who extended these features to developments for middle-income homebuyers, unfolding the "democratic utopia" described in chapter 2. The democratic utopia was also, of course, a developers' utopia, with sustained high levels of demand, plenty of land, relatively cheap capital, weak environmental regulations, and little opposition to development in the form of NIMBYism ("Not In My Backyard").

More recently, the development industry has followed the trends of other producer and service industries, with mergers and acquisitions, vertical and horizontal integration, product diversification, the deployment of new technologies, just-in-time delivery, and niche marketing, resulting in a much greater market dominance of big, publicly traded companies with complex and sophisticated operations. The profitability of smaller firms has been constrained by the economies of scale and scope enjoyed by these larger firms. Smaller firms also find it harder to deal with the dramatic increase in NIMBYism; with the widespread introduction of impact fees; and with environmental regulations that are now more complex and more strictly enforced.[5] On the other hand, neoliberal reforms that began with the Reagan administration have weakened trade unions, radically altered the system of housing finance, loosened capital markets, and weakened corporate tax law. For the larger firms, it is still a developers' utopia. The housing booms of the 1980s and the late 1990s to mid-2000s that accompanied the emergence of New Metropolitan form afforded billions in profits for the largest firms. And, although the housing market cooled down in late 2006, the long-term prospect for developers is rosy: the United States will add approximately twenty-eight million households by 2025, along with about forty-five million new jobs, turbocharging the infinite game of real estate development. Two million homes will need to be constructed each year, and non-residential construction may top three billion square feet annually. Up to $30 trillion will be spent on development between 2000 and 2025. Half the residential structures expected to be in place by 2025 did not exist in 2000. Developers' decisions in orchestrating and delivering all this will not only determine their commercial success but also influence the evolution of the form and appearance of the New Metropolis. The leading edges of the New Metropolis, in particular, are the product of the decisions of independent developers with a "supply-side aesthetic" that is heavily influenced by the market research and production decisions of the largest firms in the home building industry.

The Home Building Industry

Home building used to be mostly a local industry. Before World War II, about a third of all houses were built by small, local, general contractors for their owners, while another third were built on spec by operative builders who each averaged between three and five houses per year. Medium- and large-sized community builders accounted for the rest, though no firm produced housing for large regional markets, let alone the national market, and none had the capacity to build more than a few hundred houses a year. There are still around 80,000 home builders in the United States, most of them tiny, but the biggest builders are rapidly getting bigger and taking up an increasing share of the market. In 1986, the market share of the 100 largest builders (in terms of new home sales) stood at 24 percent. In 2006 it was 44 percent, with the top ten builders capturing just over 25 percent on their own. In a much-expanded market since the early 1990s,[6] it has been the top twenty builders who have taken market share away from the rest. Pulte Homes, the top homebuilder in 1986 with sales of 9,500, was third-largest in 2006, with sales of 41,487. The top builder in 2006 was D. R. Horton, with sales of 53,410 homes in seventy-seven markets across twenty-six states. Other top-ten firms in 2006 were Lennar Corporation, Centex Corporation, KB Home, Beazer Homes, K. Hovnanian Homes, The Ryland Group, NVR, and M.D.C. Holdings—all of them, like Horton and Pulte, publicly traded firms.[7] Between them, they accounted for the sale of almost 296,000 homes in 2006: the equivalent of the entire housing stock of a metropolitan area the size of El Paso, Texas. D. R. Horton aims to close 100,000 homes annually by 2010, while Pulte Homes aims at 70,000. If they and the rest of the top ten stay on course, they will collectively account for more than 50 percent of new home sales.

The Logic of Size

The increasing dominance of big firms is a result of a combination of factors: access to land and capital, mergers and acquisitions, geographic diversification, improved production methods, product innovation, and strategic alliances. One of the strongest driving forces in the consolidation of the building industry is the importance of access to land—not only land in the right sort of location but also in sufficient quantity for developers to be able to exercise economies of scale. Big firms have a huge advantage because it takes a lot of time to buy land and get the necessary permits to build on it, which means having the personnel for intelligence-gathering, scouting, and handling the fiscal and regulatory paperwork. Big firms also have the upper hand because they have the financial resources to acquire prime parcels of land. A $30 million deal for a piece of raw property has become commonplace, and a $150 million deal is not unheard of. With deals in this league, most large companies in fact usually purchase an option on land in the first instance rather than buying it outright. This enables the company to back out of a deal if a municipality refuses to provide the clearances needed to build.

The larger the firm, the greater the need to acquire a land bank in advance of development, simply in order to ensure a smooth flow of projects. Toll Brothers Inc. (the fourteenth-largest builder in 2006, with 8,601 closings), for example, controls enough land for nearly 80,000 houses. Its competitors, which tend to build lower-priced houses on smaller lots, have even larger land banks. Hovnanian Enterprises has enough for more than 100,000 houses; Pulte Homes holds 232,000 sites. Each vice-president at Toll who is responsible for building projects in a particular region is also charged with finding more land ("ground," in the parlance of the trade). According to Doug Yearley, a senior vice-president with responsibility for Toll developments in New Jersey, South Carolina, and several Midwestern states, "We make sure we're fully connected with all the brokers, with all the planners, with all the lawyers. . . . We let them know that we are desperately looking for more land. We put ads in the paper; we have sections in our Web site asking, 'Do you have land?' We assign territories to land-acquisition managers, and that's all they do." The employees on Yearley's land-acquisition team have to know every piece of land in every desirable town in their territory; they have to know every elected official, and have to carry around a color-coded map of every town in their territory so as to know where sewer lines and power lines run, since both can determine the viability of any potential subdivision. "Every parcel that's more than 50 acres that has not been developed," Yearley adds, "has to be identified."[8]

Big public firms have access to big capital, and usually at better rates than smaller, private firms, which have to rely on banks. Credit-rating agencies have rewarded the success of the top home builders and, coupled with the reduced risk that many builders enjoy by optioning most of their land instead of owning it, have helped the top builders secure larger credit lines at lower interest rates. The big firms have also become more creative in the ways they finance deals and manage risk, particularly with the use of joint ventures. Vertical and horizontal integration has also extended economies of scale and scope as well as intensifying the structural dominance of large firms within the industry. Lennar, for example, achieved its number two ranking in 2006 in part through its acquisition of U.S. Home in 2000, doubling the size of the company, and its subsequent purchase of twelve private builders. In 2003, Hovnanian increased its capacity by more than 20 percent through the acquisition of Summit Homes, Great Western Homes, and Windward Homes. The following year Hovnanian acquired Rocky Gorge Homes, and in 2005 Hovnanian acquired Cambridge Homes, First Home Builders of Florida, and Town and Country Homes. By 2006, Hovnanian's output was 75 percent greater than it had been in 2003. Some of the largest builders have meanwhile brought parts of their supply chain in-house in an effort to reduce cycle time (the average time it takes to complete a house) and exert more control over the tight labor supply in the construction industry.

Geographic diversification has been important to the success of big firms as well as an inevitable consequence of their growth. In 2002, D. R. Horton began using some of its existing seventy-seven markets as springboards to smaller satellite markets. The strategy—modeled after Wal-Mart's expansion across the country—enables the company to use much of the same staff and hold overhead down while capturing market share from small- and medium-size builders. It also allows them to redeploy resources from underperforming markets to stronger areas of the country—a strategy not available to smaller, single-market builders. KB Home (number five in 2006), which had concentrated almost 90 percent of its business on the West Coast in the mid-1990s, now builds in forty of the top seventy-five markets, with less than 20 percent of its business on the West Coast.[9]

The big builders have also been able to deploy new technologies and refined production methods to push down costs and undersell small and medium-size builders. Toll Brothers, for example, uses pre-fabricated wall panels and roof-truss systems, shipped from its factories to home sites; Pulte Homes has introduced prefabricated concrete foundation plates instead of site-poured concrete foundations. What is more revolutionary is that the big builders don't really build anything, at least in a technical sense: nearly all the physical work is contracted or subcontracted to electrical, framing, roofing, painting, masonry, and plumbing companies, many of which follow the big firms in itinerant fashion from development to development. These contractors do the actual building in accordance with the big firms' signature designs and management guidelines.[10]

The advantages enjoyed by big firms led to big profits during the recent housing boom. During 2005, the peak year of the boom, home prices galloped upward and so did profits. Toll Brothers, for example, increased its base prices by $3,000, $5,000 or even $10,000 every week or two. As a result, the company made almost $92,000 per home in 2005, and its 13.9 percent after-tax profit margin on the 8,769 homes it sold was the highest among all of the big publicly traded home builders, according to corporate regulatory filings with the Securities and Exchange Commission.[11]

Success Story: Pulte Homes

The firm began in 1950 when a young Bill Pulte tried his hand at building a house, having earlier worked school holidays as a construction hand. By 1956 he had incorporated his home-building activities under the name William J. Pulte, Inc., and by 1959 the firm had begun its first full subdivision, Concord Green in Bloomfield Township, Michigan. During the 1960s Pulte expanded into the Washington, D.C., Chicago, and Atlanta markets, and in 1969 the firm reincorporated through a merger with American Builders, Inc. of Colorado Springs, Colorado. The newly formed Pulte Home Corporation became a publicly owned company, allowing Pulte to extend its operations from high-priced single family homes and subdivisions of medium-priced homes to include the low-cost Federal Home Administration (FHA) and Veterans Administration (VA) housing markets. At the same time, Pulte diversified into student apartments, turnkey multifamily housing, and the "Quadrominium," a fourteen-foot wide town house designed to appeal to the baby-boomer market.

Going public prompted the firm to adopt a much more corporate approach. To control its construction costs, it implemented a computerized critical path program. To offer convenient financing and competitive mortgage rates to its homebuyers, it established the Intercontinental Mortgage Company (later renamed ICM Mortgage Corporation). Other Pulte financial services companies included Pulte Financial Companies, Inc. (PFCI), the parent company of several bond-issuing subsidiaries, and First Line Insurance Services, Inc., which provided customers (principally Pulte homebuyers) with convenient and competitively priced insurance-related services to protect themselves and their new homes. During the 1980s Pulte created a training program ("Pulte University") for entry-level construction personnel. Then, aiming at its managerial class, the firm launched the Pulte Quality Leadership (PQL) scheme. Under Pulte Quality Leadership, Pulte developed a Customer Satisfaction Measurement System and formed an in-house National Construction Council that developed performance requirements for nearly 200 distinct processes involved in building a house. The Council also implemented a comprehensive "building science" program that was the first in the industry.

These initiatives allowed the company to reexamine the entire construction process in search of competitive advantage. For example, in Charlotte, North Carolina, Pulte pioneered the completion of garage slabs, driveways, walks, stairs, and rough grading far earlier in the construction process so that realtors and brokers could show their houses to prospective customers even in bad weather. Pulte's Chesapeake operations pioneered a screw system to attach gypsum and subfloors, reducing drywall cracks, nail pops, and floor squeaks—three of the most frequently occurring problems in a new home.

The Council identified four different buyer profiles—"targeted consumer groups" for which it designed homes: the traditional family; the single person; the empty nester; and the extended family. The latter included parents with children starting college or with children in their twenties still living at home. This approach allowed Pulte to develop standardized homes that met the needs of the consumers it was targeting and reduced the need for customization of floor plans and features. Pulte developed new home designs that decreased formal areas like dining rooms and living rooms, providing space for larger kitchens with fireplaces, bigger family rooms, and master suites; and ranches gave way to two-story neoclassical revivals.

The Council also changed the strategy that Pulte used to acquire land. Instead of using the industry's traditional "price and terms" philosophy, Pulte started to select land based on an understanding of where targeted consumer groups wanted to live: paying a premium, for example, on land adjacent to large preserves of wetlands, streams, fields, and forests. More recently, the company has focused on brand identity and recognition, changing the company name to Pulte Homes, Inc., redesigning the logo, giving away a Pulte home in a national sweepstakes, and sponsoring a float in the Macy's Thanksgiving Day Parade.

By the mid-1990s, Pulte had become the largest homebuilder in the United States, with sales reaching the $2 billion mark and a portfolio that included master-planned communities and retirement communities. Its first big master-planned community, in Tampa, Florida, included 1,550 homes, a five-acre park, golf course, day care center, and a luxury swimming pool. Strategic alliances and acquisitions reinforced the company's strength. In1996 General Motors contracted with Pulte to build 6,000 homes for employees in Chihuahua and Juarez, Mexico. In 1998 Pulte acquired Radnor Homes and DiVosta, and signed an exclusive agreement with GE Appliances to supply all Pulte homes with home appliances. In 1999 Pulte Homes became a Fortune 500 company, and in 2001 the company completed a merger with the Del Webb Corporation, a high-profile specialist in master-planned developments for retirement and "active adult" communities.[12]

Following the business logic of big corporations in other sectors, Pulte has sought increased efficiency and cost savings through economies of scale, streamlining, and centralization. From manufacturers such as Toyota, Pulte has borrowed the notion of reducing product variation to save money and remove complexity. This has involved, for example, reducing the number of Pulte floor plans from 2,200 to 600, keeping only those that have proven popular from coast to coast; and choosing a common set of fixtures to use nationally. From upscale auto manufacturers, Pulte has borrowed the notion of making luxury features—such as expensive carpets and top-notch appliances, in Pulte's case—standard. This makes purchases more attractive for buyers and creates economies of scale for Pulte. Pulte's huge size has provided both imperative and opportunity to build a more efficient supply chain, using Wal-Mart's legendary supply-chain discipline as a model. Homebuilders usually let subcontractors buy materials and fixtures, but Pulte now buys directly from manufacturers in bulk, using its size to get a better price. Like Wal-Mart, Pulte uses regional distribution centers to deliver materials for a house just when they are needed, reducing the costs of maintaining a big inventory, as well as reducing at-site storage costs, damage to materials, and losses from theft. Departing from the traditional career ladder in the homebuilding sector, Pulte has recruited senior management from other sectors in order to bring in the expertise to implement and develop these strategies. These hires include a chief executive officer who was a marketer at Exxon and a logistics manager at PepsiCo, a head of customer relations from Sandcastle Resorts and Hotels, and a head of logistics from Wal-Mart.[13]

Since its inception in 1950, Pulte has built more than a half-million homes. Today, Pulte is the most geographically diverse of the big homebuilders, with projects in approximately 700 communities across twenty-seven states. In 2006 Pulte employed more than 12,400 people and earned revenues of $14.3 billion. The average price of the homes it sold in 2006 was slightly over $337,000. Just under half its sales were for first-time or first trade-up home buyers; another 19 percent were for second trade-up buyers; and the rest were buyers in "active adult" communities.

The Corporate Sprawl Machine

The development of a residential subdivision or community is not a straightforward proposition. The developer, along with other participants in the process, must constantly reposition and renegotiate as a parcel of raw land is transformed through multiple stages that involve many participants. Together, these participants form the bulk of what Harvey Molotch called the "rentier class" that dominates local politics in America. Molotch was the first to capture the shift in the political economy of place from a politics of industrial growth to a politics of real estate and property development. This shift was, of course, coincident with the deindustrialization of the economy and the suburbanization of office and retailing employment. It was paralleled by the ideological transition to neoliberalism, and by an intensification of political lobbying at state and federal levels by corporate real estate interests.

Molotch's "growth machine" thesis was that coalitions of property-based elites, in their quest to expand the local economy and accumulate wealth, drive most local politics. "The political and economic essence of virtually any given locality, in the present American context, is growth. I further argue that the desire for growth provides the key operative motivation toward consensus for members of politically mobilized local elites."[14] Molotch saw the principal agenda of these local elites—the rentier class of landowners, developers, realtors, bankers, construction companies and the auxiliary players in utility companies, engineering and technical subcontractors, retailers, chambers of commerce, local media, and so on—as securing the preconditions of growth. This involves the broad propagation of the ideology of growth and consumption as well as tactical politics around local government land-use regulation, policy, and decision making. Many political leaders become members of growth-machine coalitions because local governments rely heavily on real estate taxes to fund infrastructure and essential services such as schools, police, and fire protection. Support—often tacit support—also comes from professionals whose jobs depend on growth, including architects, engineers, landscape architects, planners, and surveyors.

The growth machine thesis has proven remarkably robust, even as local growth machines have engendered no-growth and slow-growth opposition and as the rentier class has become increasingly corporatized and centralized.[15] Urban regime theory and regulation theory have added a degree of sophistication that accommodates the broader context of political economy at the national and international levels, changing legal and institutional frameworks, and the agency of community groups. (These issues are discussed in detail in chapter 6.) The politics of growth has become more complex, more sophisticated, and more confrontational. But growth machine coalitions have meanwhile become more sophisticated and more aggressively competitive. In particular, the bigger building and development companies that have evolved over the past twenty years have more political clout and have more personnel and resources to deploy. The elite social networks of big corporate developers and their top managers, especially their contacts with big-name planners and designers, senior civil servants, and influential politicians, help them gain approvals, get regulations amended or waived, and get zoning variance requests approved. The big corporations also have the sophistication to deploy marketing strategies designed to circumvent opposition and even turn potential opponents into customers.

On the other hand, the size of the stakes involved can also lead them to exercise old-fashioned self-interest. In a *Washington Post* story on the loss of twenty-eight acres of rural land a day in Fairfax County, Virginia, as a result of new development, developer Til Hazel told reporters Glenn Frankel and Steven Fehr, "So what? The land is a resource for people to use and the issue is whether you use it well. . . . Is the goal to save green space so the other guy can look at it?" Hazel, a veteran of decades of suburban development, said, "It's a war. How else would you describe it?"[16] It is this sort of attitude that can push networking and lobbying beyond acceptable limits. The impatient director of development for North American Properties in Georgia placed a one-page advertisement in the form of an open letter in the *Gates-Chili Post* in late August 2006 in the hope of intimidating the Town Board of Chili, New York, into approving its application for a fifty-three-acre complex, to be anchored by a Target store: "In July 2006, the Planning Board declared its intention to act as lead agency on the Paul Road project. More than eight weeks later, no action has been taken by the Town to start this process. Our patience is thinning. If the Paul Road project does not move forward immediately, Chili will lose its opportunity with Target, now and in the future."[17] In this case, the town resisted the scare tactic, and the firm's director of development apologized; but power plays of one sort or another are routine in real estate development. Similarly, campaign contributions with implicit strings attached are seemingly commonplace, as are cross-memberships of corporate and public boards, dodgy-sounding insider-trading property deals, and—crossing over the legal lines—cases of bribery and petty corruption.

Development Trends in the New Metropolis

Like other entrepreneurs, developers, bankers, and mortgage managers seek to minimize risk. In terms of residential development, this conservative approach generally translates into housing for clearly established markets in which there is demonstrated spending power. Through the 1960s and 1970s this approach resulted in a preponderance of three-bedroom single-family suburban housing, reflecting the country's economic and demographic composition. In 1950, at the start of the postwar housing boom, more than half of all households had children, and the average household included 3.4 persons; single-person households accounted for slightly more than 10 percent of all households. Not surprisingly, there was little provision for atypical households—who were, effectively, excluded from new suburban tracts. Only in the 1980s, when marketing consultants caught up with social shifts that had made the "typical" household a demographic minority, did developers begin to cater to affluent singles, divorcees, retirees, and empty-nesters, adding luxury condominiums, townhouses, artists' lofts, and the like to their standard repertoire. Large builders inevitably are concerned predominantly with construction for high volume suburban development. Medium-size firms cannot afford to pay the interest on large parcels of developed land, so their preferred strategy is to maximize profits by building at high densities or by catering to the high-profit luxury end of the market. This strategy leaves small firms to use their more detailed local knowledge to scavenge for "custom" building contracts and smaller infill opportunities, at which point they will assemble the necessary materials and labor and seek to build as quickly as possible, usually aiming at the market for larger, higher-quality dwellings in neighborhoods with an established social reputation.

In the New Metropolis, market strategies are changing again in response to demographic changes and to lifestyle shifts associated with the "new economy." In 2000, only 27 percent of all suburban households were "traditional" households consisting of married couples with children. Married couples with no children made up another 29 percent, nonfamily households made up 29 percent, and "other" kinds of households made up the remaining 15 percent.[18] The average household contained 2.5 persons. The baby-boom generation that has dominated housing markets since the 1980s has been joined by Generation X households (consisting of people born between 1965 and 1979), which now compose just over half of the market for newly constructed homes. Gen-Xers carry 70 percent more debt than the baby boomers did at a comparable point in their lives, mostly because of the cost of housing and partly because of credit card debt.[19] Meanwhile, of course, baby boomers are reaching retirement years. Only 10 percent of the additional twenty-eight million households that will be formed in the United States between 2000 and 2025 will consist of households with children. By 2025 only about a quarter of all households will have children and nearly 30 percent will include only a single person.[20]

Other significant demographic changes include immigration and migration. More than 11 percent of the U.S. population in 2000 was foreign born, up from 8 percent in 1990. Although most immigrants still enter the country at gateway cities like New York, Los Angeles, and Miami, most of them move straight to where the jobs are. This means that, unlike past waves of immigrants, many of them never settle in central cities at all. The children of these immigrants, who likely will be better educated and more affluent than their parents, will surely have a big impact on suburban housing markets in the years ahead. Meanwhile, minority populations are already migrating to the suburbs in much greater numbers than ever before. According to the 2000 census, 27 percent of the suburban population in large metropolitan areas is made up of minorities, up from 19 percent in 1990. A majority of Asian Americans, half of all Hispanics, and 39 percent of African Americans live in the suburbs.[21] Migration also has a regional dimension, of course. Metropolitan regions in California, Florida, Texas, New York, Illinois, and northern Virginia have attracted large concentrations of immigrants and are distinctive for diversity in age and ethnicity. In the "new sunbelt" (Georgia, the Carolinas, Virginia, Tennessee, New Mexico, Nevada, Arizona, Colorado, Utah, Idaho, Oregon, and Washington), metro regions have attracted large numbers of African American, Hispanic, and white migrant households; while in most of the rest of the country metro regions have not attracted many regional migrants or immigrants.[22]

The relationships between employment patterns and housing locations has also been shifting. On the one hand, many people are working longer before retiring; on the other, the fluidity and variable geometry of the new economy already means that most younger people have come to expect not to remain in any job longer than three to five years. The new economy is also associated with shifting lifestyle preferences. At the top of the "hourglass economy," many households have more money but less free time than ever. Consequently, some people have grown tired of the maintenance that comes with big yards; others never want to settle into that routine to begin with. Still others are taking advantage of flexible work practices and information technology, and more people are working from home (up from 3.6 million in 1997 to 4.8 million in 2005).[23]

Nevertheless, many developers—especially small and medium-size firms with no in-house market research capability—have been slow to respond to these changing market conditions. As John T. Martin, a principal at Martin & Associates realtors, told a meeting of housing professionals at the Urban Land Institute in 2002, "Some developers still build based on 30-year-old patterns, regardless of trends. . . . Most builders are still going for 30 percent of the market [households with children]."[24] As in many other aspects of the industry, homebuilding is belatedly catching up with trends in other manufacturing industries, turning to the toolkit of market research, branding, and advertising as markets and consumer preferences change.

Consumer Preferences: The American Dream Extreme

The first thing to establish here is that the majority of Americans do still prefer to live in suburban settings. Market research by the National Association of Homebuilders puts the number of homeowners and prospective homeowners who would prefer to live in an urban setting at less than 20 percent of the total. When asked what *kind* of suburbs they prefer, the attributes that they list as important are testament to the power of the American Dream and the ideal of arcadian settings with a strong sense of community. An overwhelming majority want quiet, low-traffic settings, sidewalks on all streets, walking and biking paths, parks, natural open space, good schools, and houses that "look great from the curb." A minority (but nevertheless about one-third) say that they would be seriously influenced by the availability of public transportation.[25] These days, though, the Dream is shadowed by fear. As a result, safety and security also rank high in consumers' priorities, even in neighborhoods where there is in fact little crime. In a recent survey of Tampa homebuyers, for example, KB Home asked people what they wanted in a home. A home security system was nominated by 88 percent, 93 percent said they preferred neighborhoods with more streetlights, and 96 percent insisted on deadbolt locks or security doors.[26]

Domestic Bling.   But if the original Dream is shadowed by fear, it is also spilling over into a "Paradise Spell" of relentless individual aspiration and restless consumption.[27] Fulfillment of the recalibrated Dream means that home ownership in an arcadian setting now has to be packaged with a significant degree of suburban bling: bigness, spectacle, and affordable luxury have eclipsed mere residence. It is the American Dream Extreme. Size for its own sake evidently bespeaks affluence and status in the minds of many consumers. But size is also a necessary precondition for the long list of features that Americans, under the Paradise Spell, desire in their homes. About 85 percent of Americans want walk-in pantries. Seventy-seven percent desire separate shower stalls, 95 percent want laundry rooms, and 64 percent home offices. One in four want at least a three-car garage. Some 87 percent prefer three or more bedrooms with 44 percent wanting at least four, according to the National Association of Homebuilders.[28] Also in demand are double-story entrance areas, higher ceilings, second staircases, more—and bigger—bathrooms, bigger walk-in closets, bigger kitchens with professional-style stainless steel appliances, his-and-her dressing rooms, and extra rooms designated as teen dens, hobby rooms, sun rooms, or even just "bonus" rooms. Master bedrooms have been trumped by "master resorts," complete with exercise room, bathroom with heated floors and state-of-the-art sauna, and juice bar. Dens and television rooms have been eclipsed by home theaters. Desirable technology features include structured wiring (which allows multiple computers throughout the home to network with each other, as well as supporting multiple incoming phone lines) and smart systems that control central heating, water, television, and lighting and allow the homeowner to control everything in the house from another location. The top end of domestic bling currently includes steam showers in master bathrooms, warming drawers to keep towels toasty, sliding floor-to-ceiling bookcases that allow residents to quickly open up or close off spaces, and outdoor entertainment areas with a cabana and outdoor kitchen with a stainless-steel grill, granite countertops, a warming drawer, a wine cooler, and an ice maker.

The Paradise Spell, like the Dream, is also shadowed by fear. For the neighborhood, this means imposing gates, walls, and private security personnel. For individual homes, it means closed circuit television, sophisticated security alarms, and wireless backup (in case the landline is out). Top-of-the-line systems can also send a text message to a phone when they sense outdoor movement, enabling the homeowner to log on to the Internet to view surveillance cameras monitoring the front and back yards. In the wake of the movie *Panic Room* (2000), starring Jodie Foster, reinforced "safe rooms" accessed through secret doors were added to the desiderata of the more extremely affluent households with a nervous disposition.

Packaging and Lifestyle Enclaves. Thanks to focus groups, concept testing, surveys, segmentation research, and spreadsheet analysis, developers know the priorities of different kinds of households and can calculate almost to the dollar how much buyers are willing to pay for them. They know which households will take on a longer commute in exchange for more space, and how they have to price their products in order to get them to do so. Everything, down to carpet colors and textures, hardware on doors and cabinets, and lighting fixtures, is carefully packaged to target specific market segments. In KB Home's Tampa division, for example, managers talk about a "Mendoza line" when determining what features to include in a home. The term comes from baseball where it describes a player with a batting average hovering around the .200 mark.[29] KB Home's managers use it to describe components that are strongly desired by 70 percent or more of the home buyers in a particular market segment. "Extra closet space, a walk-in pantry and a covered patio are all above the Mendoza line, so they are included in all New River homes. People spending more than $220,000 for their home get space for a home office because it falls above the Mendoza line [for that price bracket]; more than $260,000, they get dual sinks in the master bathroom."[30]

Developers are also well aware of the tendency for affluent suburbanites to want to self-segregate into lifestyle enclaves, and routinely deploy psychographic research (e.g., on attributes relating to personality, values, attitudes, interests, and lifestyles) in order to help the process along. The developers of Ladera Ranch, in Orange County, California, for example, have used psychographics to identify "values subcultures" among potential buyers. Respondents were asked to rate how important certain things were to them—such as "making it big" or "finding your purpose in life"—whether "extremists and radicals should be banned from running for public office"; or whether they "like to experience exotic people and places." Some questions were aimed at people's feelings about "neighboring": people were asked, for instance, whether it was important that they know their neighbors, or have organized activities, or prefer privacy. Other questions were aimed at finding out whether people would pay for eco-friendly features; and so on. Four psychographic profiles emerged. The dominant group were status-conscious "Winners," people who tended to go for the glitziest, most expensive homes. There were also conservative-minded and religiously oriented "Traditionalists"; "Winners with Heart," a hybrid group of status-conscious people with a spiritual side; and "Cultural Creatives," a more liberal-minded, environmentally oriented group, less concerned with conspicuous consumption. As a result, Ladera Ranch, now home to around 16,000 people, was developed as a mosaic of neighborhoods, each tailored with distinctive attributes. One of the neighborhoods targeted at Cultural Creatives, for example, is "Terramor," where Craftsman-style houses are fitted with photovoltaic cells and bamboo flooring and the street layout has a central paseo that encourages people to get outside and socialize. For the more conservative-minded Traditionalists, the developers built "Covenant Hills," where homes have classic architecture and big family rooms, and the developers offer family-oriented activities, such as an annual Easter egg hunt.[31]

The names given to the neighborhoods in Ladera Ranch are also significant, of course. No longer are subdivisions simply named, as they were in the Sitcom Suburb era, after the developer, a famous person, a topographic feature, or an existing urban element like a crossroad or a landmark building. Developers now use a variety of tropes in their attempts to increase the perceived exclusivity of their products. Typically, the first part of the name must clearly identify that the development is not only a residential community, but also one of great distinction, located at a place with arcadian qualities. As a result, names now begin with

FIGURE 4-1. Part of the entrance to The Meadows at Olde Carpenter, a Toll Brothers development in Cary, North Carolina, under construction in 2007. (Anne-Lise Knox)

phrases like "The Estates at . . ." or "The Preserve at . . ." or "The Retreat at . . ." followed by several words to describe the former greenfield site in terms redolent of exclusivity, permanence, and community. Examples include The Preserve at Loudoun Valley Estates (in Loudoun County, Virginia), The Legends at Pocantico Hills (in Sleepy Hollow, New York), The Retreat at Carmel (in San Antonio, Texas), Enclaves at Cherry Hills (in Wildwood, Missouri), and The Reserve at Cypress Hills (in Sioux Falls, South Dakota). Unfortunately, some of the place names invented by developers and their marketers are risibly pretentious, often with gruesome affectations of spelling: Evergrene (a "sustainable" community in Palm Beach Gardens, Florida), The Meadows at Olde Carpenter (in Cary, North Carolina—see Figure 4.1), The Marque at Heritage Hunt (in Gainesville, Virginia), Myrtle Trace Grande (in Myrtle Beach, South Carolina), Sunset Pointe (in Little Elm, Texas), and Aberdeen Falls at Waterfall (in arid Las Vegas, Nevada), for example.

The Supply-Side Aesthetic: Entrepreneurial Vernacular

The residential landscapes and subdivisions of metroburbia are a product of what Dowell Myers and Elizabeth Gearin have described as the "tyranny of the minority."[32] Developers are extremely attentive to the preferences of prospective buyers of new homes, but in any given year perhaps only one percent of the population occupies newly constructed housing. Since this year's new housing becomes next year's move-up housing, the overall choices available in the housing market in any given period are formed by accretion from thin sets of preferences from previous years. The result is landscapes dominated by a supply-side aesthetic that reflect the competitive consumption of the Paradise Spell. Their very existence normalizes and consolidates the norms, values, and preferences that developers and their market researchers have divined from affluent new-home buyers, creating a moral landscape that both echoes and reproduces the aesthetic, social, cultural, political, and economic values of a relatively small but culturally dominant class fraction. So, just what has been built lately?

Building Big.   In overall terms, there is a great deal of inertia in the product range offered by the homebuilding industry. The largest builders have achieved success thanks mainly to the popularity of their subdivisions featuring single-family detached homes, and their market research reassures them that 80 percent of potential home buyers adhere to the American Dream of a single-family home in the suburbs. "Financial institutions get used to funding that same product, and the procedures for launching a big real estate investment trust require 'comparables'—assurances that a proposed development is similar to ones that have been successful in the past. Anything different from conventional suburban development doesn't get considered."[33]

Within these broad conventional parameters, though, there are some striking changes. The most remarkable statistics concern the size of new homes. The average floor area of new single-family homes in 2005 was 15 percent larger than in 1995, 36 percent larger than in 1985 and almost 150 percent larger than in 1950. The 2005 American Housing Survey found nearly 3.9 million homes in the United States with 4,000 square feet of space or more—up 35 percent since 2001. In 2005, Americans built 381,000 homes measuring 3,000 square feet or more—three times the amount in 1988, when data on homes of this size were first reported. Statistics on the number of bedrooms and bathrooms reflect the same trend: a significant bulking up, which has been accentuated in new suburban landscapes as the trend in median lot sizes has seen a reduction from 9,508 square feet in 1995 to 8,847 square feet in 2005. Even garage door sizes are getting bigger, to make room for minivans and sport utility vehicles: the typical garage bay door measured about seven feet by nine feet for years, but since around 2000 the trend is for eight- by ten-foot doors.

Big builders have been leading the trend to build big. Centex Homes has trademarked the phrases "The Most Home for Your Money" and "Big Homes. Little Prices." Toll Brothers' best seller in 2005 was their 4,800-square-foot Hampton model—1,600 square feet bigger than their best seller in 2000. The Hampton has four bedrooms and four and a half bathrooms (Figure 4.2). Its master bedroom area includes a bath with two private toilet areas and a tub on a raised platform, and offers two walk-in closets—the larger of the two is 250 square feet with cathedral ceilings and runs the length of the three-car garage below it. Available in markets from the Midwest through the Northeast, the home comes in architectural looks from Colonial and Federal to Provincial. The price starts at about $610,000.[34] In the Mid-Atlantic region, the Grand Michelangelo, the most popular plan offered by WCI Communities, of Bonita Springs, Florida, (number forty in 2006) measures 5,425 square feet. Small, custom builders are also building

FIGURE 4-2. Toll Brothers' best seller in 2005 was their 4,800-square-foot Hampton model—1,600 square feet bigger than their best seller in 2000. The Hampton has four bedrooms and four and a half bathrooms. (Anne-Lise Knox)

big as customers buy older homes—"scrapeoffs" and "teardowns"—in established, in-town neighborhoods and replace them with big, fully loaded homes: an upscale form of gentrification. Built on small lots, and by rule of thumb three times bigger than adjacent homes, these infill homes literally overshadow their neighbors. The trend has the saving grace that for every infill monster home there is one less potential addition to new sprawl.[35] But it has provoked NIMBYism and many communities have enacted "mansionization" ordinances to limit the height and size of new homes. In 2006, the National Trust for Historic Preservation identified 300 communities in thirty-three states that are experiencing significant numbers of teardowns, a jump from 100 communities in twenty states in 2002. The president of the trust, Richard Moe, called the demolition of older and historic homes a "teardown cancer" and an "orgy of irrational destruction."[36]

The emphasis in big homes everywhere, whether infill, speculative subdivision, or master-planned community, is on luxury and exclusivity. Toll Brothers has specialized in the mass luxury market since the mid-1990s, when it became known for its signature mansions (what the company called "estate homes") with cut-glass chandeliers and double-height foyers in market-tested patterns of meandering streets and cul-de-sacs with names ("The Glen at Hurley Ridge") meant to evoke afternoon teas and fox hunts. Under the Paradise Spell, the market for luxury homes has been booming. In 2005, for example, Coldwell Banker reached a total sales volume of luxury homes (priced at $1 million or above) of $55.9 billion, a 24 percent increase from 2004. The most sales were homes in the $1 million to $5 million price range, with an average price of $1.79 million.

Bigger floor plans mean bigger rooms. A review of the big builders' best-selling models in 2005 shows what it takes to keep up with the Joneses. Upstairs in most of the plans is a bonus room (usually a finished loft). Most master suites have "retreat" or "sitting" rooms in addition to dressing areas. In the master bath-room, most showers now have two or three showerheads, and most tubs are at least a foot longer than in the best-selling plans in 2000. Often, bathing has gone pub-lic: Toll Brothers' Hampton model has a raised tub in the center of its master bath-room, while WCI's Grand Michelangelo has an enclosure in the middle of its master bathroom with in-shower seating for at least four. Don't ask why. Separate, private facilities in the master bath are now standard in some top plans. Toll Brothers' top-selling two-story home in Florida, the Carrington, goes one step fur-ther, with completely separate master baths for him and her. One of Pulte's best-selling one-story designs in the San Diego area includes a "teen" room—which, five years ago, the builder called a bonus room. WCI's Grand Michelangelo includes a "keeping" room next to the breakfast area, which a spokeswoman says can be used as a gathering place for guests before a meal, or as a computer alcove.[37] The top seller for Standard Pacific in Orange County, California, comes with 300 square feet of closet space in the master suite, up from 110 square feet in their 2001 best seller. Centex's top seller in San Ramon, California, has an eight-by-six foot island (compared to a five-by-three foot island in the 2000 model) that now accommodates a vegetable sink and an auxiliary dishwasher. The stove in Standard Pacific's top-selling plan in Orange County, California, has six burners; in 2001, its most popular plan had five. And so on.

Catering to consumers' appetite for luxury not only means that rooms and floor plans have been getting bigger, but also that houses themselves must be increasingly customized in terms of appearance and offer a range of "bump-out" and "bump-up" options. A popular Toll Brothers floor plan, the Columbia (originally called the Cornell when it made its debut in 1988), can be finished with twenty-two different exterior looks, from French provincial to redbrick colonial (Figure 4.3). In this way, the company can disguise its efficiency-driven standardization, avoid the dispiriting conformity of the Sitcom Suburb era, and appeal to regional markets and consumer preferences. Hobby rooms are optional in many of the big builders' best-selling plans, as are sun rooms off the living room. Other optional extras can include secondary kitchens, private guest suites, and popular design trends such as bamboo floors and exterior glass walls. Offering a long list of options, though, can harm the builder's bottom line: in addition to the costs of carrying a big inventory, dealing with customers' different permutations of option selections can generate too many con-struction errors, delays, and costs. So firms like Pulte and Toll Brothers have whit-tled down their telephone-directory-size books of options to something more manageable, relying on market research to identify the appropriate upscale stoves, windows, fixtures, cabinets, and so on, for particular models.

Raising the Ante: Luxury Branding.   Now that well-appointed master-planned communities are raising the ante in luxury, some developers have begun

FIGURE 4-3. Toll Brothers' Columbia model (originally called the Cornell when it made its debut in 1988) can be finished with twenty-two different exterior looks, from French provincial to redbrick colonial. (The floorplans and elevations of Toll Brothers homes are copyrighted. Toll Brothers has enforced and will continue to enforce their federal copyrights to protect the investment of their homebuyers.)

to offer a new kind of option: concierge services. The developers of Creighton Farms, a 905-acre gated community in Loudon County, Virginia, have contracted with the Ritz-Carlton Hotel Company to offer residents an à la carte menu of concierge services as part of their amenity package. The developers get the exclusive branding that comes with the Ritz-Carlton name. Residents get the option of calling the property manager to have someone pick up their children from school and take them to soccer practice; or deliver a gallon of milk; or cater their dinner party and clean up afterward. The homes being built at Creighton Farms are commensurate with this kind of self-consciously over-the-top blingery. The semi-custom homes offered by WCI Communities will start at 6,000 square feet and be priced from the low $2 millions. Custom homes offered by five other builders will range from 6,000 to 13,000 square feet and start at $3 million. All will be served by an underground utilities and a fiber optic broadband network, and all the residents will be members of the golf club, with its Jack Nicklaus–designed course, year-round indoor-outdoor golf practice facility, and 40,000 square-foot clubhouse with a five-star restaurant.[38]

Niche Markets: McLofts and Master-Planned Communities

A persistent challenge for builders and developers is that many consumers are giving increasingly higher priority to luxury options and unusual features, while at the same time the market is splintering into subgroups—Gen-Xers, second-generation immigrants, empty nesters, the active elderly, and nonfamily households—with distinctive preferences: some for traditional suburbia, some for specialized or distinctive niches within suburbia or exurbia, and some for in-town settings. In response, some developers have decided to specialize in particular niche markets. The Urban Land Institute's *Residential Development Handbook* lists seven different niche markets: suburban infill, condominium conversions, "workforce housing" (a euphemism for affordable housing), adaptive use projects, seniors' housing, factory-built housing (a euphemism for trailer homes), and themed recreational communities (golf course communities, marinas, ski resorts, and equestrian communities).[39] The Urban Land Institute's InfoPacket on emerging niche markets also covers lofts, "hip" homes (e.g., Modernist designs), second homes for baby boomers, co-housing communities, developments for gays and lesbians, artists, students and soldiers, Latinos, and Generation Y (born between 1980 and 1995).[40]

Strategically, many of the larger builders have staked part of their future growth on in-town vertical construction, especially highly amenitized, high-priced condominium apartment buildings for empty-nester baby boomers and affluent Gen-Xers. A variation on this is the "McLoft": new *suburban* condominium apartment complexes, usually in the form of simulacra of nineteenth-century warehouses, with wrought-iron railings, eight-foot-tall windows, steel doors, and brick exteriors—but also with designer kitchens, bathrooms with bidets, and fiber-optic wiring. In the new Midtown Town Center in Reston, Virginia, for example, apartments priced up to $1.4 million come with old-style wide plank floors and just a few interior walls, but they also have Italian cabinets, quartz countertops, and a Morton's of Chicago (an upscale chain of steak houses) within strolling distance. In Scottsdale, Arizona, Third Avenue Lofts looks like a plain red-brick warehouse with narrow metal awnings and balconies that suggest fire escapes, while inside the building has a gym and pool, and units that come in one of thirty different floor plans and that can cost up to $4.2 million for a 3,100-square-foot apartment.

The niche-type developments with greatest impact on the landscapes of the New Metropolis, however, are private master-planned villages, neighborhoods, and communities—usually dominated by single-family housing but often containing a mixture of housing types and land uses that may include condominium apartments and townhouses as well as recreational amenities and some retail or commercial space, depending on the market niche that the developer is targeting. The result of carefully researched niche marketing and product differentiation, they offer packages of amenities and themed settings that are matched to the finances and aspirations of very specific income and lifestyle groups. Some, like Anthem, north of Phoenix, are packaged to appeal to young families. Anthem feels more like a holiday resort than a town. It includes a water park with water slides, a children's railway, hiking trails, tennis courts, a rock-climbing wall, two golf courses, several spotless parks, a supermarket, mall, two churches, a school, and a country club. Some are packaged to appeal to affluent retirees ("active adults" in the euphemism of the trade): the Del Webb development

corporation's marketers have identified early-retiring baby boomers—
"Zoomers"—as the target market for their latest Sun City development, packaged accordingly with Starbucks cafés, Internet access, and multi-gyms, as well
as the usual tennis courts, pools, and golf courses. Some are packaged as "green"
communities or "sustainable" developments and feature conservation subdivision design (CSD), attempting to preserve the visual character and ecological
functions of rural landscapes in exurbia by clustering houses together and leaving much of the site undeveloped.[41]

Others have a narrower focus: Aliante is a big master-planned development
straddling the still-to-be-completed "Beltway" on the fringe of Las Vegas on land
previously held by the Bureau of Land Management.[42] Opened in May 2003,
Aliante will take five years to build and will eventually have 7,500 homes (all
single-family homes—there is not a single multifamily or rental unit planned for
the entire development). The theme here is Nature. In fact, Aliante's address is on
Nature Park Drive. The development's signs are similar to those used by the U.S.
Park Service, and Aliante's "information gallery" looks as if it could be a U.S.
Park Service entrance building. Adjacent to the information gallery is a "Nature
Discovery Park," where the children's play section is "archeological themed" and
includes features such a "fossil wall" and "dinosaur eggs."[43] Just over a fifth of
Aliante's 428 acres is for recreational and public use, including a municipal golf
course, city parks, an arroyo, and a trail system. Aliante is technically a mixed-use
development, in that retail and offices will be built proximate to residences, but its
urban design could hardly be described as pedestrian-oriented. Homes, stores, and
businesses are all separated by walls and gates. As Lang and LeFurgy observe,
"Ironically, Aliante easily achieves the built densities—even though it has no real
multifamily dwellings—that could support a more traditional street system complete with pedestrian-accessible shopping. In fact, a person could live literally a
stone's throw (over a wall) from a store. But given the circulation system within
the development they could easily be stuck walking a mile just to reach a destination that sits no more than 50 or 100 yards away. It is more than likely that people
will simply get in their cars and drive to grab a paper or a quart of milk—something that a slight shift in urban design would have made easy to reach on foot."[44]

Ave Maria, a master-planned development twenty-five miles east of
Naples, Florida, is being developed by Domino's Pizza founder Thomas S.
Monaghan in partnership with the Barron Collier Company, an agricultural and
real estate firm. It is being built around Ave Maria University (which Monaghan
also founded) and governed by strict Catholic principles. Set on a 5,000-acre site
with a "European-inspired" town center, it encircles a massive church and what
its planners believe to be the largest crucifix in the nation, standing nearly sixty-
five feet tall. Though facing legal and constitutional challenges from civil liberty
groups, Monaghan has mandated that Ave Maria's stores won't sell porno-
graphic magazines, its pharmacies won't carry condoms or birth control pills,
and its cable television will carry no X-rated channels. He has referred to its construction as "God's will."[45]

Front Sight, just outside Las Vegas in Pahrump, Nevada, is under construction with another specialized theme. This time it's guns and ammo, featuring streets with names like Second Amendment Drive and Sense of Duty Way, target-shooting ranges, a pro shop stocked with weapons, a martial arts gym, a defensive driving track, a kindergarten through twelfth grade school where teachers will be allowed to carry concealed firearms, and sales inducements that include an Uzi machine gun and a discounted game-hunting safari in Africa.

Cooper Life at Craig Ranch, a mixed-use, fitness-based residential compound spanning fifty-four acres in the Dallas suburb of McKinney, is aimed at a very different market segment: affluent households concerned with wellness and fitness. Health consultant Kenneth H. Cooper, an authority on aerobics, has joined forces with Wellstone Communities to develop the project, with an estimated overall cost of $1 billion. Cooper Life at Craig Ranch is anchored by the 75,000-square-foot Cooper Aerobics Center and a pedestrian plaza with loft residences and retail space, and at build-out will consist of up to 700 upscale residences ranging from single-family homes to mid-rise condominiums.

Most master-planned subdivisions, though, are middle-of-the-road, family- or "active-adult"-oriented, and specialized only in terms of their "look" and the spectrum of amenities that they offer. Altogether, some forty-seven million people—almost one in six of the total population—live in one or another of the quarter million of them in the United States that have been laid out to appeal to one affluent subgroup or another. The evolution and anatomy of these developments is discussed in chapter 5.

Servitude Regimes and Regulatory Change

The common denominator among these developments is a "servitude regime": a set of covenants, controls, and restrictions (CCRs) that circumscribe people's behavior and their ability to modify their homes and yards. They are typically drafted by developers but implemented by homeowner associations, membership of which is mandatory for every homeowner in the development. The scope and detail of covenants, controls, and restrictions is numbing, covering everything from exterior trim and landscaping design to the length of time that vehicles can be left visible on the street, the weight of family dogs, the type of holiday decorations, and long lists of proscribed behaviors: putting out the laundry, making loud noise, putting up TV antennas, displaying political or religious posters, placing children's toys and equipment in the yard, and so on. Presiding over covenants, controls, and restrictions, the boards of homeowner associations amount to private governments, with power to tax residents (via homeowner dues and levies) and responsibilities not only for policing covenants, controls, and restrictions but also, in many cases, for neighborhood services such as clean-up, mowing, snow-clearing, and security. They are key to the geopolitics of metroburbia, and they are examined in detail in chapter 6.

Under U.S. law, covenants, controls, and restrictions have to be taken very seriously: if homeowner associations don't enforce them to the letter, they can be accused of being arbitrary and capricious—deadly in any court of law. So draconian measures have to be taken against even the smallest infringements (repaint those garage doors! re-lay that driveway! take down that poster!) in order to prevent the neighborhood descending into a landscape of, well, normality. The result is that the servitude regimes of private master-planned communities now define the "legal landscapes" of the American Dream. Developers like covenants, controls, and restrictions because they impose a high degree of stability on communities until they are entirely built out and sold off. While this degree of control may seem somewhat tyrannical and in violation of American ideals of individual freedom and private property rights, consumers like them because they help to enhance equity values when markets are bullish and preserve equity values when markets are flat. In the current version of the Dream, then, the accumulation of wealth trumps homeowners' freedom of individual action and expression, at least insofar as aesthetics and public comportment are concerned.

In addition to servitude regimes, two other institutional innovations have been crucial to the success and proliferation of private master-planned communities. The first is that of condominium ownership. First appearing in Puerto Rico in the 1950s and spreading to all fifty states by 1968, condominium ownership is typically associated with apartment buildings. Single-family homes are generally associated with fee simple ownership, but the community buildings, landscaping, private streets, parking areas, and recreational facilities like swimming pools and tennis courts in master-planned communities are typically held in condominium ownership. The management of condominium-owned property is the responsibility of a condominium association (a homeowner association or residential community association in the case of master-planned communities), the same entity that is responsible for the covenants, controls, and restrictions that apply to the homes that are held in fee simple ownership.

The second important innovation is the way that municipal planning has accommodated developers' need for flexibility through new approaches to land-use zoning. In particular, cluster zoning and Planned Unit Development (PUD) zoning in suburban jurisdictions, where Euclidean zoning had long been the norm, has allowed developers to offer mixed-use packages instead of uniformly residential subdivisions. Both cluster zoning and Planned Unit Development zoning apply density regulations to an entire parcel of land rather than to individual building lots. Planned Unit Development zoning additionally allows for a mix of housing types and land uses. With these zoning innovations, developers can calculate densities and profits on a project-wide basis, allowing the clustering of buildings to make room for open spaces (such as golf courses) or to preserve attractive site features (such as ponds or old barns) and facilitating a mixture of residential and non-residential elements and a mixture of housing types that can be adjusted as sales dictate. For developers, Planned Unit Development zoning offers economies of scale plus scope for product diversity and flexibility in lot sizes, street layouts, and house designs, all within a predictable regulatory framework. For planners, Planned Unit Development zoning offers the prospect of development with infrastructure and amenities provided at no cost to the taxpayer.

The successes of the product lines described in this chapter are evidence of a reenchantment of suburbia—for some, at least. Reenchanted suburbs are the immediate product of the development industry: developers' market researchers have identified the current desiderata of the "spirit of modern consumerism" among the affluent upper-middle classes (e.g., bigness, spectacle, and affordable luxury), and developers have responded with products that meet them. But these reenchanted suburbs—the themed and gated master-planned communities that dominate the leading edges of the New Metropolis—have also been mediated by architects, urban designers, and planners who have contributed to the reenchantment of suburbia through their renderings, translations, and adaptations of arcadian, utopian, environmental, and aesthetic ideals. These influences are the focus of the next chapter.

CHAPTER 5

Comfortably Numb

DEGENERATE UTOPIAS AND THEIR
EVANGELISTIC CONSULTANTS

The self-image of the design professions generally emphasizes their independent, socially progressive, environmentally sensitive, holistic, and far-sighted—if not visionary—perspectives on urban development. The practical reality, though, is a little different. Private-sector design professionals must compete for commissions and professional standing, while public-sector design professionals must strive for professional advancement within agendas set by municipal boards. Both must practice within a starkly neoliberal political economy in which progressive notions of the public interest and civil society have been eclipsed by the bottom line in corporate and public-private investment. So, for designers and planners to survive and prosper, their solutions have to be commercially attractive. Compromises have to be made, projects have to be hustled, ideas have to be sold, and unpalatable truths have to be spun into palatable propaganda. In the process, design professionals have become mythographers and fabulists, as well as social gatekeepers and shapers of the built environment.

The packaged landscapes of reenchanted suburbia bear the mark of this compromised professionalism. While they reflect and embody the professions' intellectual legacies of arcadian, utopian, communitarian, and aesthetic ideals, they do so in a way that is radically transmuted by the materialism and exclusionary impulses of their target markets and the conservatism and profit-driven competitiveness of their clients and employers. Accommodating these imperatives has drawn designers into the realm of "imagineering," which, observes Edward Relph, "has become one of the primary ways of making landscapes."[1] David Harvey, borrowing from Louis Marin's categorization of Disneyland, has described the packaged landscapes of master-planned communities as paradigmatic "degenerate utopias."[2] Like Disneyland, they are designed as harmonious and nonconflictual spaces, set aside from the "real" world. Like Disneyland, they incorporate spectacle and maintain security and exclusion through surveillance, walls, and gates; and, like Disneyland's Main Street, they deploy a sanitized and

mythologized past in invoking identity and community. All of this is "degenerate," in Harvey's view, because the oppositional force implicit in the progressive
and utopian ideals embraced by the design professions has slid, in the course of
materialization, into a perpetuation of the fetish of commodity culture. This chapter explores the roles of design professionals as key actors in the creation of the
reenchanted suburbs and degenerate utopias of the New Metropolis.

Evangelistic Bureaucrats and Mythographers

Urban design professionals find themselves in a pivotal and problematic
position, inextricably implicated in the successes and failures of urban development. Yet their roles are often ambiguous and ambivalent. They cast themselves
as stewards of the built environment and guardians of the public interest, and are
perceived that way by large sections of the public. At the same time, they are servants of their clients and employers. They do not legislate policies, nor do they,
typically, invest capital in urban development. Yet their influence on outcomes is
critical: a "triumph of the eunuch."[3]

The Janus-faced condition of the urban design professions—to take a somewhat less anatomical metaphor—is founded on a deep paradox that persists in the
make-up of contemporary planning and urban design. The paradox was this:
Although urbanization was the vehicle that industrial capitalism needed in order to
marshal goods and labor efficiently, it created dangerous conditions that threatened public health and resulted in crowded settings within which the losers and the
exploited might organize, consolidate, and rebel. Urban design and planning, as a
response to this paradox, were born as hybrid creatures, dedicated at one level to
utopian ideals of humanistic and sanitary reform, but charged on another with
ensuring the efficient management of urban land, infrastructure, and services, as
well as fostering and maintaining "balanced" and stable communities. As such,
planning and urban design should be construed as key to the internal survival
mechanisms of capitalism, transforming the energy of opposing social forces into
the defense of the status quo and helping the dominant order to propagate its own
goals and values as the legitimate ones.[4] Their practitioners can be seen as
Weberian "ideal types": key actors in the development process, with distinctive
values and ideals that are routinely transcribed into the built environment through
their day-to-day work as well as through their more visionary prescriptions.

Design professionals are in fact a relatively narrowly defined, self-selected
group whose values and ideals are condensed and socialized through formal education, professional meetings, trade journals, and career reward systems.[5] Formal
education in the design professions has always been informed and inspired by the
bold and utopian ideas of Ebenezer Howard, Patrick Geddes, Frederick Law
Olmsted, Daniel Burnham, Le Corbusier, Frank Lloyd Wright, and Clarence
Stein; and by tales of a modern professional history featuring can-do pioneers as
well as coffee-table intellectuals and celebrity designers.[6] From these roots there
developed an evangelical spirit to the entire profession: Cities *could* be better

places; they *should* be. In this spirit, the founding ideology of urban design and planning was an amalgam of the liberal idealism, utopianism, communitarianism, and environmentalism of the early reformers, philanthropists, and urban design-ers, along with a concern with aesthetics that stemmed from links with architec-ture. The influence of European Modernists and the Congrès International d'Architecture Moderne (1933) and the subsequent publication of its proceedings in the Charter of Athens (1942) added a penchant for rationalistic, sweeping, futuristic solutions. After World War II the design professions engaged with the latest developments in social science research and theory, exploring and acquir-ing the languages and toolkits of behavioral theory, regional economics, regional science, quantitative geography, systems analysis, and transportation modeling.

The result, when combined with their evangelical spirit, was a professional make-up and disposition that proved tragically unsuited to the ideals that they espoused. Even in their finest hour—before egalitarian liberalism was edged aside by neoliberalism—they were forced to watch themselves fail. Their evan-gelism and environmental determinism led them to pursue "bureaucratic offen-sives" like urban renewal, sometimes to the point where the communities whose lives they had hoped to improve were angry and afraid. In the suburbs, they became technicians and enforcers of Euclidean zoning and drafters of subdivision layouts and regulations.[7] Before they knew it, their rationality and their predilec-tion for efficient and tidy land use patterns had led them to become deeply impli-cated in most of the disenchanting aspects of suburban sprawl.

Architects, by and large, escaped criticism of this sort, largely because they remained aloof from all things suburban. Domestic architecture had become pro-fessionally unfashionable after the demise of the Bauhaus in the late 1930s. In the democratic utopia of patternbook designs in the 1950s and 1960s, it was also unprofitable. Exceptions were made, of course, for "cover-shot" commissions for homes for the rich and/or famous. Meanwhile, losing interest in the suburbs meant relinquishing professional claims on urban design. In universities, urban design programs became academic backwaters and have only recently begun to recover vitality.

For most of the second half of the twentieth century, professional architects looked to accumulate fees from large public and commercial projects, while profes-sional magazines celebrated innovation, spectacle, and "starchitecture"; and the star architects themselves competed to build (or at least draw) in radical and spectacu-lar fashion.[8] Peter Eisenman, Rem Koolhaas, Zaha Hadid, and Daniel Liebeskind, for example, gained notoriety in the 1980s for "deconstructivist" buildings that fea-tured nonrectilinear shapes, improbably intersecting volumes, and geometric imbalance. It coincided with interest in the humanities and social sciences in Deconstructivist philosophy, but had no real connection except that the opaque writings of Jacques Derrida were handy in lending an apparent intellectual weight to the fashion. The ideology and professional self-image of architecture was domi-nated by aesthetic formalism of one stripe or another, with leading practitioners relying on an ability to remain one step ahead of the capacity of their audience to understand or critique their work.[9] Little attention was given in the architectural press, the professional magazines, or schools of architecture to metropolitan development or domestic architecture. The result, to quote scholar-practitioner Bernard Tschumi, was that architecture became characterized by "disjunctions and dissociations between use, form, and social values."[10] When attention did turn to broader issues of metropolitan development, it was mostly to tut-tut about the shortcomings of placeless "Generica" and the formulaic outcomes of suburban building.[11]

Eventually, as big but unglamorous firms began to get profitable commissions from developers for office parks, suburban town centers, and master-planned communities, more attention was paid within the profession as a whole to the challenges of design for what had hitherto been considered mundane settings. A battle against the mutations of sprawl became a worthy (and potentially profitable) task for architecture. Schools began to revive urban design programs, and new academic journals began to stake out the intellectual territory.[12] Practitioners took the lead in advocating transit-oriented development, neotraditional design, and a "new urbanism" that aims to codify the fundamental physical principles that embody region, neighborhood, and community. Planners, meanwhile, have seen an opportunity to rehabilitate both their profession and the suburbs by reviving an interest in urban design and reclaiming the intellectual heritage of utopian and communitarian ideals.[13] It is an interesting but unnerving mix: planners are not really trained to undertake (or understand) design, while the complex and recursive dynamics of the economic, social, cultural, and political dimensions of urbanization are generally a closed book to architects.

It is not surprising, therefore, that their collective efforts have found expression in the degenerate utopias of the New Metropolis. The key antecedents of these degenerate utopias, however, go deep into the history of urban design, including the philosophical roots of the American Renaissance and various strands of reactionary praxis as well as the influence of landmark projects such as the "restorative utopias" of the late nineteenth and early twentieth centuries and the "new community" experiments of the 1960s.

Antecedents: Restorative Utopias, New Communities, and Reactionary Excursions

The intellectuals' utopias of the nineteenth century (chapter 2) found expression in garden suburbs like Riverside, Highland Park, Lake Forest, and Winnetka, and garden cities like Forest Hills Gardens: "restorative utopias" that would cater not only to the upper-middle classes but to the full spectrum of society, with jobs and civic amenities as well as homes. The typical architectural styles of these garden suburbs and garden cities were vernacular revivals with a rustic look, echoing the arcadian ideology of the intellectuals' utopias. Clarence Perry's conviction that the layout of a project could, if handled correctly, foster "neighborhood spirit," added an important communitarian dimension to the early conventional wisdom of urban design. Other foundational legacies include the ideal of settings in which man and Nature could achieve a state of balance amid pastoral and picturesque settings and the conviction, as articulated by Raymond Unwin and Barry Parker, that the central mission of urban design is to promote beauty and amenity; that creativity in urban design comes from an imaginative understanding of the past. Patrick Geddes added the concept of the "natural region" as a framework within which to reconstruct social and political life, and introduced the idea of the "valley section," an idealized transect across natural regions that locates different land uses according to their ecological niche in relation to an urban center. The City Beautiful movement subsequently introduced the aspiration of creating moral and social order through the arrangement and symbolism of the built environment.

These legacies were all carried forward as the design professions struggled to cope with the challenges of "automobility" and its attendant propensity for sprawl.

Master Suburbs and Planned Communities

In the mid-1920s there were several attempts to plan communities based on the new demands and possibilities of automobility. The best-known of these are Palos Verdes Estates (Los Angeles), Shaker Heights (Cleveland), River Oaks (Houston), the Country Club District (Kansas City), and two "master suburbs" in Florida: Coral Gables and Boca Raton. While each had its own innovations, they were all built by private developers for profit, with an upper-middle-class, automobile-owning market in mind. They were characterized by very low densities (for the time) of about three dwellings per acre, by high-quality landscaping, by the inclusion of recreational facilities (golf courses, in particular), public gardens, and plazas in addition to shopping amenities, and by detailed deed covenants aimed at preserving the character and appearance of the entire development.[14]

In retrospect, the Country Club District was perhaps the single most influential of these. It was the creation of developer Jesse Clyde Nichols, who later founded the Urban Land Institute, an independent research organization concerned with urban land use and development from the developers' point of view. Nichols had been impressed by the garden city movement and by the City Beautiful movement, and he was determined to put together a project large enough to sustain a self-contained community. Nichols had to be concerned with profitability, however, so that from the start he set out to create a setting that would appeal to the most lucrative section of the residential market: the top end. It took him fourteen years to acquire the land he needed, before embarking, in 1922, on the construction of 6,000 homes and 160 apartment buildings that eventually housed over 35,000 residents.

The centerpiece of the development was Country Club Plaza, the world's first automobile-based shopping center. Built in Spanish colonial revival style (inspired by the Spanish colonial revival style of the 1915 Pan-Pacific Exposition in San Diego), it featured waterfalls, fountains, flowers, and expensive landscaping, with extensive parking lots behind ornamental brick walls. Nichols carefully controlled the composition of businesses in the Plaza through leasing policies that brought upscale retail stores to the first floor of the development and lawyers, physicians, and accountants to the offices on the second floor. Similarly, the residential sections of the Country Club District were carefully landscaped and controlled. Densities were kept low, streets were curvilinear, trees were preserved wherever possible, and houses were set back some distance from the street, with driveways and garages. All sales were subject to racially restrictive deed covenants, and all purchasers were required to join the homeowners' association, the purpose of which was to ensure lawn care and to supervise the general upkeep and tidiness of streets and open spaces. It was a commercial success from the very start and, despite the obvious elitism of the whole venture, it also attracted critical acclaim from the developers and planners who came from across the country to view the shape of the future.

The Regional Planning Association of America (RPAA) also grappled with the implications of automobility. While Regional Planning Association of America leaders like Lewis Mumford were mostly interested in big-picture, long-term scenarios for urbanization and regional development, they did manage, through the agency of Alexander Bing—a New York real estate developer and founding member of the Regional Planning Association of America—to create two planned communities. The first, Sunnyside Gardens (built between 1924 and 1928), was an undeveloped inner-city site in Queens, only five miles from Manhattan. Here, Clarence Stein and Henry Wright designed big traffic-free superblocks that enabled the creation of vast interior garden spaces. The second, and most significant as an influence on later master-planned communities, was Radburn (started in 1928), fifteen miles from Manhattan in the borough of Fair Lawn, New Jersey, where Stein and Wright were able to release the Sunnyside superblock principle from the rigid grid of the inner city. Traffic was channeled through a hierarchy of roads, so that most residential areas could be kept virtually free of traffic. Pedestrians and cyclists were given their own paths that crossed traffic arteries under rustic bridges, while housing was clustered cozily around irregular-shaped open spaces. The Regional Planning Association of America had hoped to draw in a socially mixed group of residents, but the attractiveness of Radburn quickly ensured that it became a commuter suburb for upper-middle-class families, whereupon realtors took it upon themselves to keep out Jews and African Americans in order to maintain property prices. Nevertheless, Radburn, like the Country Club District, became an influential landmark in the history of urban design.

The New Community Experiment

In the quarter century after World War II there were few exceptions to Sitcom Suburbs. In the Chicago area, communities such as Park Forest, Elk Grove, and Oak Brook were distinctive for having generous extra measures of green space within an otherwise standard 1950s format of curvilinear streets, culs-de-sac, and shopping strips. Elsewhere, some developers and design professionals proposed large-scale, private master-planned communities. "Ranging in projected population from ten thousand to five hundred thousand, these communities were planned to be phased, coordinated, socially balanced, environmentally aware, and economically efficient. Their developers wanted to create whole communities rather than simple subdivisions. By avoiding many of the problems of uncoordinated incremental growth—or sprawl—they imagined both improving urban areas and creating a real estate product that would sell."[15] This was the "new community experiment." It was inspired by the success of government-sponsored postwar new town programs in Britain, France, and Scandinavia, and it carried forward an eclectic mix of influences, including Perry's neighborhood unit concept and the arcadian ideals of the nineteenth-century seers.

Two of the most notable examples of these new communities were in the Washington, D.C., area. Columbia, Maryland, and Reston, Virginia, were planned suburbs that were organized into "villages" of 10,000 to 15,000 people, each with a village center, recreational facilities, and schools. Meadows and woods separated the villages from one another and from a system of open space corridors containing walkways and bicycle paths. Reston is noted for its emphasis on aesthetics. Reston's layout gave strong consideration to the preservation of the site's natural topography and woodland. Built form was intended to create a distinctive sense of place: the first village to be developed, Lake Anne, was loosely modeled on the Italian fishing village of Portofino, complete with apartments and shops enclosing a waterfront piazza. Columbia was organized into nine villages, each with a mixed-use village center consisting of stores, schools, and community facilities. Columbia's developer, James Rouse, placed special emphasis on creating socially diverse and nonsegregated communities. The cellular village structure common to these and other new communities such as the Irvine Ranch, in Orange County, California, and The Woodlands, north of Houston, Texas, was not only an application of Perry's neighborhood unit concept but also a framework intended to create settings with an identifiable sense of place and to foster social interaction, civic engagement, and a sense of community. For the developers, the cellular structure was especially attractive because it provided a neat, logical way to phase the build-out of what were planned to be unusually large developments, requiring substantial investments of capital.

Ann Forsyth, in her detailed review of Columbia, the Irvine Ranch, and The Woodlands, argues that each has made significant contributions to urban design, including mixed land use and housing types, walkable neighborhoods, distinctive identities and sense of place, and open space preservation.[16] In comparison with the incremental subdivision development of the 1960s and 1970s, planned new communities were found by researchers at the University of North Carolina to be significantly better in terms of accessibility and the quality of community and recreational facilities, but no different in terms of residents' evaluations of housing and neighborhood livability, rates of participation in neighboring, community organizations, community politics, the organization and operation of community governance, and satisfaction with life as a whole.[17]

Reactionary Movements

While the new communities experiment was an attempt on the part of a few developers and design practitioners to provide a profitable and attractive alternative to the standard tract development of the era, others were reacting in a more radical way to the shortcomings of rationalized, large-scale, suburban sprawl. Much of this reaction was in the form of "paper architecture," a mix of theory and advocacy on the part of writers and aspiring boutique practitioners, with only isolated examples of building.[18] In the 1950s a "Townscape Movement" emerged in Britain as a reaction to the sculptural architecture of Modernism and the lack of urbanity and human scale in the British New Towns. The movement stressed the "art of relationship" among elements of the urban landscape, and the desirability of the "recovery of place" through unfolding sequences of street scenes, buildings that enclose intimate public spaces, and variety and idiosyncrasy in built form. These were sentiments that were echoed in the United States by influential public intellectuals Paul Goodman and Jane Jacobs. Richard Sennett, an academic who was on his way to becoming a public intellectual, wrote elegantly and persuasively about the importance of public space and urban design in the social life of cities.[19]

Kevin Lynch, on the planning faculty at the Massachusetts Institute of Technology, introduced the notion of the legibility of townscapes and explored how to harness human perception of the physical form of cities and regions as the conceptual basis for good urban design. Christopher Alexander, an English architect who taught at Berkeley, sought to identify a "language" of patterns among elements of built form and public spaces, the rationale being that a knowledge of such patterns might be useful in imbuing urban design with "timeless" sensibilities. The innovative methodologies deployed by both Lynch and Alexander were naïve and unreliable, and their logic was based on raw environmental determinism (built-form stimulus → human/social/cultural response); but their results were unquestioned and their work was influential, largely because it lent an analytic dimension to a spreading concern for the qualitative aspects of urban landscapes.[20]

A different strand of reaction to the loss of urbanity came from the neo-rationalist movement. The most influential of the early neo-rationalists was the Italian scholar-practitioner Aldo Rossi. His book *Architecture of the City*, published in 1966, was a critique of what he saw as Modernism's denial of the inherent complexity of cities. As an alternative to the totalizing models of Modernist architecture, the neo-rationalists sought to identify various "types" of architecture appropriate to economic and geographic context. Neo-rationalists saw the built environment as a "theater of memory" and hoped to identify "the fundamental types of habitat: the street, the arcade, the square, the yard, the quarter, the colonnade, the avenue, the centre, the nucleus, the crown, the radius, the knot. . . . So that the city can be walked through. So that it becomes a text again."[21] During the 1970s these ideas were pursued by the Movement for the Reconstruction of the European City, with Léon Krier as its chief exponent. Krier was an advocate of urban design based on identifiable, functionally integrated quarters, and of architecture with the proportions, morphology, and craftsmanship of the pre-industrial era, with set-piece ensembles of buildings (as in neoclassical Bath, Berlin, and St. Petersburg). He was influenced by Tönnies's sociology and the conviction that small-scale towns provide the preconditions for *Gemeinschaft*, the most intense form of community.[22] Neo-rationalism had less impact in the United States than in Europe, partly because the conduit for neo-rationalist ideas in the United States was an elite and self-referential group: Peter Eisenman, Michael Graves, Charles Gwathmey, John Hejduk, and Richard Meier. They were known as the New York Neo-Rationalist School but, with the singular pretension that is a common affliction among architects, they liked to be called "The Five" or "The Whites." They were concerned with the design of single-family homes, but mainly from the point of view of pure aesthetics, unclouded by social or cultural concerns (hence the "Whites").

The placelessness of modernist development also produced some more straightforward reactions. Some argued for a neo-classicism, looking to classical architecture for inspiration and seeking to recover precedents and typologies from past reasonings about spatial order. Others argued for regionalism and vernacular design. Lewis Mumford, following Patrick Geddes, had argued in the 1920s for the preservation of regional architectural traditions. In 1964, the Museum of Modern Art in New York put on an exhibition entitled *Architecture Without Architects*, which greatly stimulated an interest in vernacular architecture. The curator of the exhibition, Bernard Rudofsky, urged architects to look to "non-pedigreed" architecture for inspiration.[23]

In France, the impulse to regenerate traditional urban qualities resulted in "Provincial Urbanism." The goal of Provincial Urbanism was to create "the traces, the arrangements of streets and plazas, the types of housing (and especially individual houses with yards), the perspectives, the source of architectural composition which made our cities so pleasing, particularly our provincial cities before being submerged, first by the growth of suburban tract developments, then by the brutal push of the 'grand ensembles' with their density and severe geometry."[24] Several developments in this vein gained international attention. One was Port Grimaud, designed and developed by François Spoerry near Saint Tropez in 1973 to resemble a fisherman's village. Also influential was the first architectural exhibition of the Paris Biennale, in 1980, which took the theme "In Search of Urbanity." The organizer of the exhibition, Jean Nouvel, saw urbanity in terms of the attributes of places that express and nurture their inhabitants' lifestyles and aspirations, reflecting identity and memory, conflict and change. In Italy, the inaugural international architectural exhibition of the Venice Biennale, also in 1980, took the theme of "The Presence of the Past: The End of Prohibition," seeking to recast urban design theory by "reawakening the imaginary." One way or another, history and geography were reentering the discourse on urban design.

In the United States, the advocacy of "contextualism" was based on a similar desire to recognize the legacy of past development. Contextualism placed special emphasis on drawing upon all the elements at hand, including the existing fabric, and stressed the importance of the street, the axis, and the role of building mass as a definer of urban space. Its principal advocate was Colin Rowe, who used the metaphor of a collage to emphasize the idea of bringing together diverse elements.[25] Taking this sentiment a lot further, Robert Venturi and Denise Scott Brown urged architects and urban designers to "learn from Las Vegas" and the creative energy of the contemporary vernacular of commerce and advertising. Theirs was a particularly strong reaction to the totalizing rationalism of modernity with its goal of purity, unity, and order. They argued for the exact opposite: messy vitality, hybridity, ambiguity, and inconsistency.[26] In doing so, they unleashed a virulent episode of "postmodern" architecture—though they were quick to disown the results.

The intellectual basis of postmodern architecture bore little relation to that of postmodern theory in the social sciences and humanities, but it did reflect and parallel a broader shift in cultural sensibilities that abandons modernity's emphasis on economic and scientific progress, focusing instead on living for the moment. Above all, postmodernity is consumption-oriented, with an emphasis on the possession of particular combinations of things and on the style of consumption. Postmodern society is a "society of the spectacle" in which the symbolic properties of places and material possessions assume unprecedented importance.[27] Since the mid-1970s, postmodernity has permeated every sphere of creative activity, including art, architecture, advertising, philosophy, clothing design, interior design, music, cinema, novels, television, and urban design.

With these reactionary influences playing in the background, with developers driven by economies of scale and inspired by imagineering, and with consumers with an appetite for bigness, spectacle, and affordable luxury, it was perhaps inevitable that the reenchantment of suburbia should take the form of private, master-planned developments redolent with historic motifs.

Instant Karma, Packaged and Gated

For places to have good karma for the affluent upper-middle classes, they have to have status, amenity, security, and a sense of community as well as a distinctive identity. Private, master-planned communities confer status and amenity through their self-consciously expensive homes packaged with an array of exclusive features, amenities, and services. Gates—real or implied—and perimeter walls and fences also confer status, even though security is often cited as the motivating force behind the trends in gated communities (especially when they are "forted up" with gatehouses, security patrols, and surveillance). In the suburbs, gates are there mainly for show. Ed Blakely and Mary Snyder make a broad distinction between "lifestyle communities" (such as retirement communities, golf communities, and suburban new towns), "prestige communities" (enclaves of exclusive housing) and "security zone communities."[28] Security is the primary concern only in the latter—neighborhoods where residents have retrofitted gates or barricades in order to fend off or regain control from some outside threat: street crime, more often than not.[29]

Lifestyle communities and prestige communities are gated more for status than security: "Used to giving out their gate codes to plumbers and pizza delivery boys, residents have few illusions about the security of their enclaves."[30] Ironically, research by Setha Low suggests that living behind gates in lifestyle- and prestige-type communities can actually intensify people's fear of the unknown in the rest of the city.[31] Nevertheless, the *appearance* of security confers an aura of status and exclusivity, while gates and walls also help to control the social environment, effectively restricting certain groups and so segregating the community by class and race. "Unlike real estate covenants of the past, gated community restrictions are legal and are rarely challenged for being discriminatory."[32] In addition, by excluding unwanted persons or activities, gates are an effective way to protect property values. Gates are a common feature in the subdivisions and master-planned communities of metroburbia. In 2005, almost seven million households lived in communities surrounded by walls or fences and more than four million lived in communities where access was controlled by some means.[33]

A sense of community—or, at least, the promise of a sense of community—is also part of the package with private master-planned communities. At one level, a sense of community derives from the secession and congregation of self-selected and narrowly defined class, income, and lifestyle groups. At another, a sense of community, along with a sense of place, is carefully cultivated by developers through the theming and branding of their developments, the names that they give them, and the advertising copy they deploy. The commodification of community and sense of place has become central to residential development. As Blakely and Snyder observe, "Developers use 'community' as a term of art to discuss their products in promotional materials. Marketing brochures are written to convey a sense of community, referring to housing tracts as 'new communities within a city,' 'a totally new way of life,' 'an old community setting,' or even 'your new hometown.' "[34] Developers' most effective tropes in packaging community and sense of place, though, derive from the use of neotraditional design and the exploitation of the brand recognition of new urbanism.

Life in the Past Lane

Writing in the early 1990s, Vincent Scully asserted that "the most important movement in architecture today is the revival of vernacular and classical traditions and their reintegration into the mainstream of modern architecture in its fundamental aspect: the structure of communities, the building of towns."[35] Coming from America's most venerable architectural historian, the assertion is compelling. The movement initially found expression in what proponents called Traditional Neighborhood Development (TND), an attempt to codify tract development in such a way as to create the look and feel of small-town, pre–World War II settings in which pedestrian movement and social interaction are privileged over automobile use. "In the place of 'pods' of housing 'clusters,' office 'parks,' and shopping 'centers' assembled along 'collector roads,' the Traditional Neighborhood Development is based on grids of straight streets and boulevards (instead of highways) which are lined by buildings in order to generate clear and enclosed public spaces. Buildings are grouped by scale and architectural expression but house a variety of functions, social classes, and age groups."[36] Traditional Neighborhood Developments generally avoid having culs-de-sac because they are thought to inhibit social contact. Front driveways and garages are proscribed because they are felt to be ugly, to affirm the primacy of the automobile, and to be out of keeping with traditional and vernacular house types. Architects Andres Duany and Elizabeth Plater-Zyberk are generally regarded as the progenitors of Traditional Neighborhood Development, and the details of architecture are an important element of the approach. A "traditional" small-town neighborhood flavor is typically pursued through design codes that result in housing that mimics pre–World War II housing styles, relegating garages to the back lot, restoring street-focused front porches, and placing houses on small lots. In a similarly motivated attempt to provide guidelines for a new typology of suburban development, architect Peter Calthorpe developed the concept of the "Pedestrian Pocket." Harking back to the days of streetcar suburbs, Calthorpe's idea was for higher-density suburbs to be situated within a quarter-mile walking distance of public transportation hubs, ideally light railroad stations. Thus, pedestrian pockets would become part of a regional scheme of "Transit-Oriented Development," or TOD.

Further codification of ideas about neotraditional urban design emerged from an exercise sponsored by California's Local Government Commission. Peter Katz, a Commission staff member and an advocate of neotraditional urban design, brought together a group of architects that included Calthorpe as well as Duany, Plater-Zyberk, and Traditional Neighborhood Development advocates Stefanos Polyzoides, Elizabeth Moule, and Michael Corbett. They were asked to come to agreement about a set of community design and development principles in order to provide a vision for an alternative to urban sprawl. Their findings were presented in the fall of 1991 to an audience of local elected officials at a conference at the Ahwahnee Hotel in Yosemite. The resulting "Ahwahnee Principles"[37] set out a series of progressive objectives that emphasize sustainability and sense of place in urban design. Reasonable, and at face value mom-and-apple-pie virtuous, they are nevertheless rather myopic in that they take virtually no account of the imperatives of developers or the preferences of consumers. The principles do not explicitly refer to neotraditional design, but they have had a significant impact within the design professions in that they have provided a declarative platform and a vision, based on the essentialized characteristics of prewar, small-town America, that has rekindled a spirit of evangelism.

Meanwhile, the developers of upscale subdivisions and master-planned communities have been busy responding to the *zeitgeist* by building homes with nostalgic and vernacular references. Throughout the East Coast, neocolonial and neo-federal references are most popular (Figure 5.1). By one estimate, about 1.7 million homes were built in neotraditional style in the United States in 2006.[38] Not constrained by any sense of authenticity, and safely assuming that "buyers can't tell the difference between Craftsman and Mediterranean,"[39] the imagineered results are often monstrous in appearance as well as size. Appearance

FIGURE 5-1. East Coast neotraditional: these homes in Brier Creek, Raleigh, North Carolina, are redolent with neocolonial and neo-federal references, including carriage lamps, colonnaded entranceways, balustrades, fanlights with multiple keystones, double-hung sash windows, splayed brick lintels with keystones, and (fake) window shutters. (Anne-Lise Knox)

notwithstanding, the names given to upscale models also draw on the resonance of heritage and nostalgia. Pulte's line, for example, features the "Legend," the "Tradition," and the "Legacy." WCI Homes has a "Thoreau" model. More wide-spread is the tendency to draw freely on old-world place names, regardless of any association (or lack of it) between the style references embodied by the model and the vernacular or period styles actually associated with those places. Lennar, for example, has the "Windsor" and the "Dorchester"; Pulte has the "Chatsworth," the "Hastings," and the "Aberdeen"; WCI has the "Devonshire" and the "Canterbury"; and Toll Brothers has the "Southport" and the "Keswick."

The preferred styling for signage, street furniture, and nonresidential build-ings in many upscale subdivisions and master-planned communities—especially on the East Coast—also draws heavily on nostalgic and vernacular references and is a major selling point. The advertising copy for Murray Hill Square in New Jersey, for example, points out that "each distinctive home is a one-of-a-kind reproduction of Colonial and Victorian landmarks . . . set in a fairy-tale-like, turn-of-the-century village, complete with brick-lined courtyards, formal box-wood gardens, gaslights, and village squares."[40] The Withers Preserve, near Myrtle Beach, South Carolina, is billed as giving residents "the unique ability to enjoy the benefits of an active lifestyle with easy access to shopping, dining and enter-tainment while still living in a small-town neighborhood atmosphere. Quality homes, designed in traditional Lowcountry style, will be available in a variety of models."[41] The Reserve at Cypress Hills, South Dakota, offers "A Planned

Community. . . . Architecture that evokes styles of yesterdays craftsmen and style worthy of preservation."[42]

Nostalgia and the Appeal of Traditional Form

The emergence of nostalgia as such a strong element in the contemporary *zeitgeist* is part of the broad economic and cultural shifts of the late twentieth century. The economic system shock triggered by the quadrupling of crude oil prices by the Organization of the Petroleum Exporting Countries cartel in 1973 called into question the longstanding assumption—inherent to modernity—that the future would bring scientific progress and increased prosperity. It is no accident that this attenuation of faith in the future coincided with a revaluation of the past and the growth of a "heritage industry" with multiple dimensions. In cities, one important dimension was the conservation and preservation of the historic fabric. Between 1970 and 1985 the number of properties and districts listed in the National Register of Historic Places increased from 1,500 to 37,000. By 2007 the number had reached 98,528.

Economic and cultural globalization, intensified by the economic system-shock of the mid-1970s, was another factor. The more universal the diffusion of material culture and lifestyles, the more local and ethnic identities are valued. The faster the information highway takes people into cyberspace, the more they feel the need for a subjective setting—a specific place or community—they can call their own. The faster their neighborhoods and towns acquire the same generic supermarkets, gas stations, shopping malls, industrial estates, office parks, and suburban subdivisions, the more people feel the need for enclaves of familiarity, centeredness, and identity. The United Nations Centre for Human Settlements (UNCHS) notes, "In many localitiés, people are overwhelmed by changes in their traditional cultural, spiritual, and social values and norms and by the introduction of a cult of consumerism intrinsic to the process of globalization. In the rebound, many localities have rediscovered the 'culture of place' by stressing their own identity, their own roots, their own culture and values and the importance of their own neighbourhood, area, vicinity, or town."[43]

The impulse for nostalgia resonates well with the cultural turn to postmodernity and the associated eclecticism, playfulness, historicism, pastiche, and, above all, spectacle.[44] Guy Debord argued that spectacle has become the overarching dimension of consumer society, including the packaging, promotion, and display of commodities and the production and effects of all media.[45] This shift to a society of the spectacle involves a commodification of previously noncolonized sectors of everyday social life. In this vein, Douglas Kellner points out that in addition to the obviously spectacular elements of the built environment—waterfront redevelopments, "festival" settings, theme parks, and signature buildings like the Frank Gehry Guggenheim Museum in Bilbao, Spain, and the Richard Meier Getty Center in Los Angeles, for example—there are also more mundane elements "that embody contemporary society's basic values and serve to enculturate individuals into its way of life" and so become defining phenomena of their era.[46] Kellner calls these phenomena megaspectacles (though metaspectacles may be a more appropriate term) and cites McDonaldization as an example. Neotraditional landscapes in packaged and themed master-planned communities would clearly be another.

Some neotraditionalists in the design professions see historic references and vernacular form as filling the "emptiness" of contemporary urban life, or as an element of coherence in the face of splintering urbanism. Others see neotraditional design as instilling a moral landscape that is conducive to "community." It has also been suggested that neotraditional settings provide a stable point of reference for residents whose pace of life is frenetic.[47] Robert Yaro, now president of the Regional Plan Association, has endorsed traditional neighborhood development on the grounds that genuinely historic settings have become "too expensive for most people."[48] Peter Calthorpe, meanwhile, has effectively blamed consumers for neotraditional design: "I would posit that in my experience as a practitioner, most of the neotraditional style comes from the marketplace itself, not from the intentions of any designers or an intentional design ethos."[49]

Design critics, on the other hand, characterize neotraditional settings as mawkish; infantilized Disneyfication; camp architectural costume drama; backlot sets based on a decontextualized past or on fictional histories; and as "hyperreal" environments based on cultural reductiveness.[50] Nan Ellin refers to the nostalgic reflex to "drag and drop forms from other places and other times" as "form following fiction."[51] Architecture theorist Kenneth Frampton, in an early reaction to the glibness of the nostalgic reflex in architecture, called for a more "critical regionalism" that might assimilate genuine local materials, crafts, topographies, and climate with the broader trends of national and global culture. At the heart of these critiques is the question of authenticity, a key concept in contemporary sociology and cultural anthropology. As Martin Heidegger anticipated, the "authenticity" of place has been subverted as people's capacity for "dwelling" has been attenuated by the effects of rationalism, mass production, standardization, and mass values.[52] Jean Baudrillard famously asserted the dominance of simulacra and simulation in contemporary society, sparking a debate as to whether authenticity is (or ever was) an innate condition of people or places. "Authenticity" turns out to be a very slippery and contestable concept, open to abuse and misinterpretation.[53]

"Tradition," on close interrogation, is equally elusive. Echoing Edward Said, Janet Abu-Lughod has warned of the use and abuse of the notion of "tradition" to reinforce or maintain established forms of domination. In other words, the idea of tradition is a rhetorical device, and "traditions" themselves can be seen as being identified, manufactured, packaged, and deployed in pursuit of social inclusion and/or exclusion.[54] From this perspective, neotraditional architecture and traditional neighborhood development are inherently socially regressive. Richard Sennett, for example, describes them as "exercises in withdrawal from a complex world, deploying self-consciously 'traditional' architecture that bespeaks a mythic communal coherence and shared identity in the past." He describes their designers as "artists of claustrophobia" and concludes that "place making based on exclusion, sameness, and nostalgia is socially poisonous and psychologically useless: a self weighted with its insufficiencies cannot lift the burden by retreat into fantasy."[55] This perspective is consonant with the idea of neotraditional landscapes as metaspectacle. For Debord, the spectacle is a tool of pacification and depoliticization; it is a "permanent opium war" that stupefies social subjects and distracts them from the most urgent task of real life—recovering the full range of their human powers through creative practice. In submissively consuming spectacles, people are estranged from actively producing their lives.

Déjà vu Urbanism

Undeterred by such critique, the adherents of neotraditional design sought to seize the initiative by rebranding their ideas and principles under the banner of "new urbanism" (*the* New Urbanism—capitalized—to its adherents). A meeting held in Alexandria, Virginia, in 1993 was proclaimed as the first Congress for the New Urbanism (CNU). Andres Duany—by all accounts a master of evangelical oratory and rhetoric—became the principal spokesperson, enthusiastically assisted by supporters like consultant Peter Katz and journalist James Kunstler.[56] Annual meetings—congresses—attracted considerable interest among design professionals and by 1996 a charter had been ratified. Echoing the Ahwahnee Principles, the charter is a list of what seem, at face value, rather reasonable and progressive desiderata about urban and regional development and redevelopment.[57]

Aggressively evangelical, well-organized, and media-savvy, new urbanists became a sort of Salvation Army for the built environment. The CNU was established as a nonprofit organization (with both individual and institutional membership) and adopted an idealistic stance founded on the credo—hardly new—that there is such a thing as "good" urbanism and that it can be propagated through the codification of design principles. The principles are based on precedents and typologies derived from observations of patterns exhibited in traditional communities. Citing the enduring popularity of walkable, diverse, urban atmospheres in places like Nantucket, Rhode Island; Alexandria, Virginia; Georgetown, D.C.; historic Charleston, South Carolina; and Savannah, Georgia, new urbanists place special emphasis on a traditional vocabulary of urban design, with a typology that includes boulevards, perimeter blocks, plazas, monuments, and the pedestrian scale of streets and public spaces. The canon was established by Duany and Plater-Zyberk, whose firm, DPZ, drew up a "Lexicon of New Urbanism" and shared it with the Congress for the New Urbanism. The physical configuration of streets is key to new urbanism, as is the role of building mass as a definer of urban space, the need for clear patterns among elements of built form and public spaces, and the importance of having identifiable, functionally integrated quarters.

The belief is that both civic architecture and pedestrian-oriented streets can act as catalysts of sociability and community. Tree-lined streets are designed with a comparatively narrow width, and lined with stoops or front porches as social buffer zones between the public realm of the street and the private realm of the home. As in Traditional Neighborhood Developments, culs-de-sac are avoided; small lots, mixed uses, and side alleys are encouraged. Towns are conceived as being made up of a series of clearly identifiable neighborhoods and districts, with pedestrian-oriented commercial enterprises and civic spaces like schools, parks and community centers distributed throughout the neighborhoods, and vehicular traffic routed through boulevards that provide axes of orientation. Larger commercial activities are concentrated in a town center, along with significant civic structures such as churches and local government buildings. As a counter to the sprawl and splintering urbanism of the New Metropolis, each neighborhood, district, and town should have clear centers and edges, as in Geddes's idealized transect of a "natural" region. The regional scale is also an important dimension of new urbanism, allowing towns and their surrounding countryside to be integrated into a sustainable planning framework that can address issues of air and water quality, transportation, equity, diversity, and sprawl.

All this is to be achieved, according to the Congress for the New Urbanism, through a sort of painting-by-numbers for urban designers. Detailed prescriptive codes and conventions, embedded in a series of regulatory documents—a Regulating Plan, Urban Regulations, Architectural Regulations, Street Types, and Landscape Regulations—provide the template for new urbanist developments. In addition, a "SmartCode," developed and copyrighted by DPZ, allows them to be zoned incrementally along the lines of the urban-rural transect. Douglas Kelbaugh, dean of the Taubman College of Architecture and Urban Planning at the University of Michigan and a proponent of new urbanism, describes the approach as utopian, inspirational, structuralist, and normative:

> It is utopian because it aspires to a social ethic that builds new or repairs existing communities in ways that equitably mix people of different income, ethnicity, race and age, and to a civic ideal that coherently mixes land of different uses and buildings of different types. It is inspirational because it sponsors public architecture and public space that attempts to make citizens feel they are part, even proud, of a culture that is more significant than their individual, private worlds and an ecology that is vertically and horizontally connected to natural loops, cycles and chains. . . . It is structuralist (or at least determinist) in the sense that it maintains that there is a direct, structural relationship between physical form and social behavior. It is normative in that it posits that good design can have a measurably positive effect on sense of place and community, which it holds are essential to a healthy, sustainable society.[58]

If the principles of new urbanism sound (to borrow from Yogi Berra) like déjà vu all over again, it is because there is almost nothing that is new about new urbanism apart from its sophistication in organization, branding, and marketing. While the immediate antecedents of new urbanism are Traditional Neighborhood Development and the Ahawahnee Principles, it is substantively a derivative hybrid of ideas and impulses that go back to the intellectuals' utopias of the nineteenth century and that incorporate elements of the City Beautiful; Nolen's association of urban design with the tenets of classical civic ideals; Geddes's natural region and urban-rural transect; Clarence Perry's neighborhood unit idea; Raymond Unwin and Barry Parker's assertion of traditional and vernacular design; the precedents of the garden suburbs of the late nineteenth century, the master suburbs of 1920s, and the new communities of the 1960s and 1970s; the British Townscape Movement; Richard Sennett's ideas on importance of public space; Christopher Alexander's notion of pattern language; Kevin Lynch's concept of legibility; and the prescriptions and inclinations of neo-rationalism, Provincial Urbanism, Contextualism, and Postmodernity.

What the new urbanists have done is to craft a manifesto that contains a great deal of this accumulated wisdom and made it seem appropriate to the moment. New urbanism appeals to public officials because, as Jill Grant observes, its rules for practice "allow planners to operate without worrying about the messiness of divergent claims."[59] New urbanism appeals to developers because its closely written codes alleviate many of the problems of customer choice associated with new suburban development and their associated complications and increased costs. They also like it because it has a strong brand identity with sensibilities—a vision of community and a sense of certainty, respectability, and predictability—that appeal to the key market segment of upper-middle-class households. The brand identity of new urbanism is strong for several reasons. One is that the Congress for the New Urbanism has been successful in mobilizing large quantities of resource materials and propaganda, both in traditional form—coffee-table books, magazine articles—and on the web, with news updates, image banks, press kits, reports, a project database, and a store with books, DVDs, posters, and T-shirts. Another is the Congress for the New Urbanism's organization, with regional chapters in addition to a national framework with international connections, and thematic "salons" featuring "conversations" on design, education, environmental issues, planning, and transportation. In addition, Duany himself has been a charismatic and insistent figurehead, spokesperson, and salesman for what is, after all, not only "his" movement but also his business.

By the late 1990s, new urbanism had made allies in the press, captured the imagination of a section of the public, and gained the attention of the Urban Land Institute and the U.S. Department of Housing and Urban Development. In 1998 the National Town Builders Association formed to help developers interested in new urbanism. The movement received a steady stream of coverage in professional magazines, and in 2002 the American Planning Association established a new urbanism interest group.[60] It was the "triumph of the eunuch" redux. New urbanism has certainly enlivened interest in planning and urban design, and brought fresh ideas to what had become routinized and bureaucratized issues of land use and zoning. It has also reinforced sense of place, livability, sustainability, and quality of life as important policy issues, and helped to resurrect the idea of a definable public interest.

Degenerate Utopias: The Icons.   Exhibit A in the case for new urbanism has been the resort town of Seaside, Florida, one of DPZ's first projects, established in 1982.[61] Laid out with a central square, a grid street plan modified with radial-concentric boulevards recalling City Beautiful and Garden City principles; an urban code controlling the interdependency between road width, landscaping, lot size, and housing type; and an architectural code drawing on southern vernacular houses in pastel colors, Seaside was very photogenic. Although it was an isolated resort town on Florida's Gulf coast, it quickly became an icon of new urbanism. By the mid-1990s it had become financially very successful. Ironically, its notoriety and market appeal drove up prices to the point where it became a seasonal and second-home resort for the very affluent, thus undermining any potential for fulfilling new urbanist claims with regard to any kind of genuine urbanism. Nevertheless, it provided, at the moment of the formal inception of the Congress for the New Urbanism, a real-life model of what a new urbanist development might look like.

Another early icon for new urbanists was Laguna West, near Sacramento, California, designed by Peter Calthorpe. Begun in 1990, Calthorpe's plan called for a walkable development of 2,300 units with a commercial center in traditional style, eventually to be linked to Sacramento via rapid transit. Marketing and financial pressures, however, meant that the development was modified as it was implemented, resulting in a subdivision that is more scenic and at a higher density than conventional developments, but otherwise similar to other California suburbs, with residents living car-dependent lives. Back in Florida, the Disney Corporation broke ground on a new town, Celebration, twenty miles southwest of Orlando, in 1994, just as the Congress for the New Urbanism was forming.[62] Although not formally new urbanists, Celebration's designers, Robert Stern and Jaquelin Robertson, planned Celebration according to new urbanist principles, and the town has subsequently become cited as another examplar, for better or worse, of new urbanism. Celebration has a grid plan for more than 8,000 residential units and a town center with more than two million square feet of retail space in a mixed-use town center that includes apartments above stores, a school, a branch college campus, and a hotel as well as office space. There is an imposing town hall with twenty-eight columns and a gigantic door, but no town government—the town manager is a Disney executive.

Architectural conformity in Celebration is ensured by a seventy-page pattern book of house designs inspired by the kinds of places featured in *Southern Living* magazine. Curtains that face the street must be white or off-white. The color of a home, unless it is white, must not be duplicated within three homes on the same side of the street. At least a quarter of the front and side gardens must have some vegetation besides grass; and so on. The town's many "traditions" have been imagineered by the Disney Corporation, in true Disney style. These include snow every night from the day after Thanksgiving to New Year's Eve— the "snow" consisting of soap bubbles. For two weekends in October, oak leaves (fabricated from tissue paper) fall from palm trees downtown. The net result is "a picture-book town with a Kodak moment on every corner, and it reminds all first-time visitors of a film set."[63] Retailing in the town center is heavily oriented to tourism, so that everyday shopping requires residents to get in their cars and drive to the nearest mall.

Kentlands, in Gaithersburg, Maryland, is generally regarded as the most successful of the commonly cited new urbanist developments. Another design from DPZ, Kentlands has a mix of housing types from detached single-family homes and townhouses to condominiums and apartments; a town center with a church, a school, and a community center; and a Main Street with restaurants and storefront offices that screen a regional mall from the residential areas. Kentlands is well endowed with sidewalks, a series of beautifully landscaped small parks, and plenty of playgrounds, pools, basketball courts, and clubhouses. Residents can walk from almost anywhere in the 354-acre, 1,700-home development to shop at a Whole Foods grocery store, get a coffee or a meal, drop off dry cleaning, or visit the doctor. Surveys of residents have elicited relatively high levels of satisfaction with their environs, but critics have pointed out that, its mix of housing types notwithstanding, Kentlands is an enclave of affluent upper-middle-class households.[64]

Between them, these icons of new urbanism have inspired hundreds of others. *New Urban News* began to track new urbanist projects being developed in the United States in 1999, when it recognized 124 of fifteen acres or more. By 2003, the number was up to 636. After that, the brand identity of new urbanism had become so powerful in the marketplace that it became impossible to make an accurate count: many developers of master-planned communities were putting a "new urbanism" label on projects whose packaged landscapes include the "look" of new urbanism—neotraditional house styles, extra parks and sidewalks, and a fake town hall, perhaps—but that fell well short of the prescriptive codes endorsed by the Congress for the New Urbanism. Take, for example, Avellino, a development near Windermere, Florida, announced in 2007 by KB Home. Its principal brand identity will come from a tie-in with domestic diva Martha Stewart, whose marketability has evidently rebounded since she was released from federal prison in March 2005, having served five months for lying about a 2001 stock sale. Avellino's houses will feature architectural styles inspired by Stewart, including a number of units "themed after some of her own homes," according to George Glance, president of KB Home's Orlando division. In an interview with the *Orlando Sentinel*, Glance noted that the subdivision will have a "new urbanism look," with front porches, garages at the rear of the homes accessible by alleys, and smaller lots with community open spaces and sidewalks to encourage people to walk and mingle in the neighborhood; but it will not include any retail space or "town center."[65]

The Critique: Naiveté and Determinism.   In spite of its commercial appeal, new urbanism has come in for a great deal of criticism and is seen by many as jejune and meretricious. Its practitioners and advocates are portrayed as poignantly confused: an architectural *derrière garde*, trading on antique truisms that have been naively combined across time and space to form a New Age urbanism that is part conventional wisdom and part fuzzy poetic, resonant but meaningless. New urbanists only have themselves to blame for much of the criticism: their evangelical zeal has frequently carried over into the bombast and hyperbole of zealotry; or alternatively into a complacent vanity of bald-faced assertions.

More substantively, Alex Marshall, echoing Jane Jacobs's admonition that "a city cannot be a work of art," has critiqued new urbanism on the grounds that its proponents do not take into consideration how cities actually work: that real urbanism emerges from a city- or metropolitan-wide physical infrastructure and political economy. In particular, developments of 5,000 or 10,000 people cannot support economically viable "town centers" that will adequately serve the needs of their populace. "This point has always confused architects," he observes.[66] From this perspective, new urbanism can be seen as a fetishized, strapped-on kind of urbanism, segregated and isolated from the rest of the metropolitan fabric. Marshall also points to the design shortcomings of new urbanist codes, such as impractical restrictions on accessibility for automobiles and garbage pick-up as a result of displacing parking and truck-borne deliveries to side- and back alleys. Others point out that the antiseptic products of form-based codes preclude the close-grained diversity and unexpected encounters—both visual and social—that are the true glory of cities and a fundamental component of any real urbanism.

The principal underlying weakness of new urbanism is one that is shared not only with all of its antecedents, but also with modernist designers: the conceit of environmental determinism and the privileging of spatial form over social process. As Edward Robbins notes, "The New Urbanism, like modernism, can be accused of a kind of essentialism, in which all aspects of the complex and diverse urban world is reduced to a set of singular and authoritative principles summarized in a set of simple statements and strategic visual and verbal discourses."[67] In the prescriptive reasoning of new urbanism, this essentialism is magnified and laid bare. Design codes become behavior codes; cultural myopia masquerades as universal values. Central to the new urbanist pitch is that well-fashioned design codes will be conducive to community life and that specific design fetishes—front porches, narrower streets, and smaller lot sizes, for example—will actively produce community by bringing neighbors together more closely and more frequently, and therefore they will be inclined to greet and chat with each other more often.[68] Good (i.e., new urbanist) design equals community, civility, and sense of place; bad design equals placelessness, ennui, and deviant behavior.

This, of course, is a chimera: place is socially constructed, and the relationships between people and their environments are complex, reflexive, and recursive. David Harvey asks *what kind* of community is understood within the philosophy of new urbanism. Paul Clarke provides an answer: "It is an ambiguous and unformulated ideal, part romantic, part spiritual, part utopian, part pragmatic, and part illusory. . . . New Urbanism's utopian ideal of community presumes a monolithic, collective memory and is premised upon a peculiarly American mythology, a nostalgia for a past landscape that was never so picturesque and seldom without discord."[69] Harvey points out that "the darker side of this communitarianism remains unstated: . . . 'Community' has ever been one of the key sites of social control and surveillance, bordering on overt social repression. Well-founded communities often exclude, define themselves against others, erect all sorts of keep-out signs (if not tangible walls)." It is exactly because of this, of course, that new urbanism has such market appeal for developers of master-planned communities. Like the City Beautiful, new urbanism manages to match maximum planning with maximum speculation. And like the City Beautiful, new urbanism rests on the authoritarian and regressive aspiration of creating moral and social order through the arrangement and symbolism of the built environment.

Other items on the charge sheet against new urbanism include its intrinsic bias in favor of the middlebrow sumptuary codes of upper-middle-class fractions; that it adopts the rhetoric of sustainability even as it promotes an agenda of growth (for without growth, where would the consultants' fees come from?); that it adopts the rhetoric of social diversity and affordability even as it deliberately caters to a "mix" of residents with above-average incomes; that (like many of its antecedents) it is fundamentally anti-urban; that its adherents and practitioners place a premium on civility but show no interest in changing power relations; that they are anti-intellectual, disinterested in theory-building or the empirical validity of their ideas; and that their thinking is dominated by simplistic dichotomies.[70]

Exposure to any of this critique unleashes an indignant response among new urbanists. The response of the Congress for the New Urbanism itself is unyielding. It has become a sort of politburo with its own New Speak "in which subdivisions become towns, lots with smaller backyards are an antidote to sprawl, owners of real estate metamorphose into town fathers, homeowners associations supplant local governments, zoning is terrible but codes are good."[71] On the principle that the best defense is offense, the Congress for the New Urbanism has propagated a hagiographic and shamelessly self-referential literature, with a good deal of solemn nonsense contributed by naive journalists and credulous practitioners. Together, prisoners of denial and victims of aporia, they have conjured a consensual hallucination that amounts to a kind of architectural and theoretical parallel universe that avoids contact with evidence or analysis (except developers' bottom-line analyses).

Jill Grant points to another response to criticism: the addition of "theory" to bolster new urbanist rhetoric. In its early days, new urbanists were disinterested in theory. Calthorpe even argued—nonsensically—that "because the social linkages are complex, the practical must come first."[72] Don't just stand there, do something! But the weight of criticism has induced new urbanists and their sympathizers to invoke aspects of theory. Hence the disinterment, for example, of Geddes's transect concept and its associated ideas of environmental and social ecology. Grant observes, "As is common in planning theory and practice gener-ally, the new urbanists pick and choose elements of theory from diverse sources. Thus they can claim both Camillo Sitte and Raymond Unwin as inspirations while simultaneously dismissing many of the ideas for which Sitte and Unwin became famous."[73] Emily Talen, a planning academic with strong sympathies for new urbanism, has invoked sociobiology and neurobiology as theoretical props for the prescriptions of new urbanism.[74] Nevertheless, armchair philoso-phizing notwithstanding, new urbanism remains largely atheoretical, so that it cannot be "wrong" in any strictly scientific sense.

What new urbanism has evidently got right, along with other kinds of master-planned developments, is its market appeal. As premium spaces designed to accommodate the secession of the successful in enclaves that are legally sequestered by servitude regimes, they are perfectly suited to the shift in social, cultural, and political sensibilities that has occurred with the rise of neoliberal-ism. The next chapter examines the implications of this shift for the social and political geography of the New Metropolis.

CHAPTER 6

The Politics of Privatism

The artful fragments of suburban and exurban development that constitute the newest residential fabric of the New Metropolis are the seedbeds of a new politics that finds expression at several geographic scales. In terms of community governance, the degenerate utopias of master-planned developments have been characterized as fragmented "privatopias," in which "the dominant ideology is privatism; where contract law is the supreme authority; where property rights and property values are the focus of community life; and where homogeneity, exclusiveness, and exclusion are the foundation of social organization."[1] Privatopias are premium spaces designed to accommodate the secession of the successful in enclaves that are legally sequestered by servitude regimes of covenants, codes, and restrictions. They are culturally hermetic spaces, "purified" arenas of social reproduction, dominated by material consumption and social segregation. Administered as common interest communities by homeowner associations, private master-planned communities have an internal politics characterized by unprecedented issues of control, democracy, citizenship, and conflict resolution.

Meanwhile, at the meso-scale of metropolitan politics, the fragmented social geography of common interest communities and suburban and exurban public jurisdictions makes for a balkanized political landscape, animated by NIMBYism and growth/slow-growth/no-growth disputes. This new politics has significant implications for issues of governance, property rights, the public interest, and civil society. It has to be seen in the context of the broader transformation of the political economy of the United States and its metropolises, not least the shift in political sensibilities that has occurred with the eclipse of egalitarian liberalism by neoliberalism.

Spaces of Neoliberalization: Policy and Planning for the Private Interest

The ascendance of the free-market doctrines of neoliberalism has been a circular and cumulative process. The failure of Keynesianism (the operational policy framework for egalitarian liberalism) to cope with the economic system-shock of the sudden quadrupling of crude oil prices by the Organization of the Petroleum Exporting Countries in 1973, the consequent overaccumulation crisis, and the subsequent globalization of industrial production all opened the way for radically different policy perspectives. Increased taxation (to fund spending on the casualties of deindustrialization), unemployment, and inner-city decline contributed to resentment among more affluent sections of the taxpaying public, who were caught up in an ever-escalating material culture and wanted more disposable income for their own private consumption. With pressure on public spending, the quality of public services, public goods, and physical infrastructures inevitably deteriorated, which in turn added even more pressure for those with money to spend it privately. People's concern to have their children attend "good" schools intensified demand for housing in upscale developments with their own community pre-schools and elementary schools. Increasing numbers of people began to buy private security systems,[2] enroll their children in private extracurricular lessons and activities, and spend time at the mall rather than the local playground. It is only human nature that people paying for private services will tend to resent paying for public services that they feel they no longer need. Also resentful of continued spending on socially and geographically redistributive programs, they began to support the view of certain policy experts and politicians who were demanding "fiscal equivalence"—where people and businesses "get what they pay for."

The concept of the public good was tarred with the same brush as Keynesianism, as government itself (to paraphrase Ronald Reagan) came to be identified as the problem rather than the solution. Whereas market failures had been the rationale for the ascendance of egalitarian liberalism, government failures became the rationale for neoliberalism. Globalization also played a part: Keynesian economic policies and redistributive programs came to be seen as an impediment to international competitiveness. Labor-market "flexibility" became the new conventional wisdom. Republican Party leaders like Senator Trent Lott developed a populist line that played to these trends: "You know how to spend your money better than the government." Thanks to the composition and dynamics of Republican politics in the 1980s and 1990s, this economic fundamentalism became inextricably linked with a moralizing social conservatism, producing the peculiar mix of conservatism and libertarianism that has become the hallmark of contemporary America. By the mid-1990s, neoliberalism had become the conventional economic wisdom, even among mainstream Democrats. For example, John O. Norquist, a Democrat and mayor of Milwaukee for sixteen years, is credited with having helped to "reinvent" municipal government, taking on labor unions, introducing private competition for city services, and advocating school choice and welfare reform. Norquist is also a new urbanist—he became president of the Congress for the New Urbanism in 2003 in an explicit move to signal the organization's commitment to free-market libertarian ideals and the belief that development is best left to developers, property owners, and private consultants.

As Jamie Peck, Adam Tickell, and others have pointed out, all this is part of a continuous process of political-economic change, not simply a set of policy outcomes.[3] Peck and Tickell have characterized the process in terms of a combination of "roll-back" neoliberalization and "roll-out" neoliberalization. Roll-back neoliberalization has meant the deregulation of finance and industry, the demise of public housing programs, the privatization of public space, cutbacks in redistributive welfare programs like food stamps, the shedding of many of the traditional roles of federal and local governments as mediators and regulators, curbs on the power and influence of public institutions like the labor unions and the U.S. Department of Housing and Urban Development, and a reduction of investment in the physical infrastructure of roads, bridges, and public utilities. Roll-out neoliberalization has meant "right-to-work" legislation, the establishment of public-private partnerships, the development of workfare requirements, the assertion of private property rights, the encouragement of inner-city gentrification, the creation of free-trade zones, enterprise zones, and other deregulated spaces, the assertion of the principle of "highest and best use" for land-use planning decisions, and the privatization of government services. So complete is the contracting of services in some cases that small municipalities operate with only a handful of full-time employees. Weston, Florida—a city of nearly 70,000 people—has just three employees, while Sandy Springs, Georgia, an Atlanta-area baby boomburb of more than 80,000 residents, has only four public employees who are not involved with public safety. Except for police and fire, virtually every government function has been contracted out.[4] Meanwhile, with neoliberalism established as an ideological "commonsense," it was a short step to what Neil Smith has called revanchism: reclaiming urban spaces by displacing and excluding homeless and low-income people through coercive legal and police force in the cause of a "good business climate."[5]

The net effect has been to "hollow out" the capacity of the federal government while forcing municipal governments to become increasingly entrepreneurial in pursuit of jobs and revenues; increasingly pro-business in terms of their expenditures; and increasingly oriented to the kind of planning that keeps property values high. Brenner and Theodore suggest that the implicit goal of neoliberalization at the metropolitan scale is "to mobilize city space as an arena both for market-oriented economic growth and for elite consumption practices."[6] Indeed, the proponents of neoliberal policies have advocated free markets as the ideal condition not only for economic organization, but also for political and social life. Ideal for some, of course. Free markets have generated uneven relationships among places and regions, the inevitable result being an intensification of economic inequality at every scale, from the neighborhood to the nation state. The pursuit of neoliberal policies and free market ideals has dismantled a great deal of the framework for city building and community development that Western societies used to take for granted: everything from broad concepts such as the public good to the nuts and bolts of the regulatory environment.

Uncivil Society: "You Take Care of Your Own"

Inevitably, the ascendance of neoliberalism has affected the tenor and vitality of civil society. The meaning of civil society has changed over time but is generally understood to involve all the main elements of society outside of government—the "parapolitical" elements that serve as mediating agencies between individuals and the formal machinery of institutional politics. These include business organizations, trades unions, homeowner associations, and voluntary groups of all kinds, such as charities and conservation societies. Although relatively few such organizations are explicitly "political" in nature, many of them are politi*cized* inasmuch as they occasionally pursue group activities or campaigns through the medium of government. Indeed, there is a school of thought among political scientists which argues that private groups are highly influential in raising and defining issues for public debate.[7] According to this school of thought, politicians and officials tend to back off until it is clear what the alignment of groups on any particular issue will be and whether any official decision making will be required. In essence, this gives municipal governments the role of umpiring the struggle among private and partial interests, leaving them to decide the outcome of major issues in all but a formal sense.

Understanding the changing dynamics of the parapolitical structure, as Jürgen Habermas pointed out, requires a working-through of the relationship of the public to the private spheres of life.[8] At core, this relationship rests on the recognition of individual rights, freedoms of speech, and the rights that are related to the transactions of private owners of property. The way these rights are articulated and upheld in particular locales determines, among other things, the nature of access to economic and political power and to social and cultural legitimacy. At issue here, of course, is the role of homeowner associations as part of the parapolitical structure. As Ed Blakely and Mary Gail Snyder have noted, the trend toward privatized government and neighborhoods is part of the more general trend of splintering urbanism in the New Metropolis, "and the resulting loss of connection and social contact is narrowing the bonds of mutual responsibility and the social contract." They go on to observe:

> Almost imperceptibly, the societal idea of what it means to be a resident of a community seems to have changed; it is more common now to speak of taxpayers than of citizens. . . . In gated communities and other privatized enclaves, the local community that many residents identify with is the one within the gates. Their homeowner association dues are like taxes; and their responsibility to their community, such as it is, ends at that gate. . . . One city official in Plano, Texas, summed up his view of the attitude of the gated community residents in his town: 'I took care of my responsibility, I'm safe in here, I've got my guard gate; I've paid my [homeowner association] dues, and I'm responsible for my streets. Therefore, I have no responsibility for the commonweal, because you take care of your own.'[9]

Setha Low suggests that common interest communities, with weak social ties and diffuse interpersonal associations among homogenous populations, promote what M. P. Baumgartner calls "moral minimalism"—a reluctance to get personally involved in any kind of political dispute. Only when residents can be assured that someone else will bear the burden of moral authority, enabling them to remain anonymous and uninvolved, are they likely to participate in any kind of overt exercise of social control.[10]

Private Governance: The Stealthy Tyranny of Homeowner Associations

In contrast to the relative passivity of some elements of the parapolitical structure, common interest communities and homeowner associations (also known as residential community associations) have come to represent an increasingly important element within the parapolitical structure.[11] Legally, these are simply nonprofit business organizations that are established to regulate or manage a residential subdivision or condominium development. In practice, they constitute a form of private government whose rules, financial practices, and other decisions can be a powerful force in local politics. Through boards of directors elected by a group of homeowners, they levy taxes (through assessments), control and regulate the physical environment (through servitude regimes of covenants, controls, and restrictions attached to each home's deed), enact development controls, maintain commonly owned amenities (such as meeting rooms, exercise centers, racquetball courts, and picnic areas), and organize service delivery (such as garbage collection, water and sewer services, street maintenance, snow removal, and neighborhood security).

The rationale for homeowner associations presiding over restrictive covenants designed to act as instruments of "mandated permanent perfection" can be traced back to the mid-nineteenth century railway suburb of Llewellyn Park in West Orange, New Jersey, to Ebenezer Howard's garden city idea at the end of the nineteenth century, and to Clarence Perry's concept of the neighborhood unit in the first decades of the twentieth century. The earliest homeowner associations, from the first examples in the 1920s to the point in the mid-1960s when a new wave of suburbanization provided the platform for the proliferation of a new breed of associations, were chiefly directed toward exclusionary segregation. They were, as Mike Davis put it, "overwhelmingly concerned with the establishment of what Robert Fishman has called 'bourgeois utopia': that is, with the creation of racially and economically homogeneous residential enclaves glorifying the single-family home."[12] Their activities involved crude and straightforward legal-spatial tactics. At first, the most popular instrument was the racially restrictive covenant. This was a response to the Supreme Court's judgment against segregation ordinances enacted by public municipalities (*Buchanan v. Warley*, 1917); it was, in turn, outlawed by a Supreme Court case (*Shelley v. Kraemer*, 1948). Later, they turned to campaigns for incorporation that would enable them, in their metamorphosis to public governments, to deploy "fiscal zoning" (e.g., limiting the construction of multifamily dwellings, raising the minimum lot size of new housing) as a means of enhancing residential exclusivity.

Evan McKenzie suggests that it was Radburn, New Jersey (1928), based on the Progressive city manager model and with a set of covenants, controls, and restrictions (CCRs) drawn up by attorney and political scientist Charles Ascher, that became the key prototype for today's common-interest communities.[13] The Radburn model of community governance was refined and extended in the new community experiments of the 1960s—Irvine, Reston, Columbia, and others. Thus the homeowner associations of today's master-planned communities "are Americanized, third-generation descendants of Howard's utopian garden city idea."[14] They are a pervasive element of the New Metropolis, with a sophisticated institutional support network in the form of the national Community Associations Institute, based in Alexandria, Virginia. The private nature of these associations means that they are an unusually "stealthy" element of the parapolitical structure. Altogether, it is estimated that there are over 286,000 homeowner associations in the United States (compared with fewer than 500 in the early 1960s and around 20,000 in the mid-1970s), together covering more than 20 percent of the nation's households and fifty-seven million people. According to the Community Associations Institute, four out of five U.S. homes built since 2000 have been in homeowner association-governed subdivisions. At least half of all housing currently on the market in the fifty largest metropolitan areas and nearly all new residential development in California, Florida, New York, Texas, and suburban Washington, D.C., is subject to mandatory governance by a homeowner association. In Arizona, Pima County alone has more than 800 associations representing about 100,000 homeowners.

As described in chapter 4, the explosive growth of homeowner associations in recent years has been driven by the logic of the real estate industry, which sees mandatory membership in pre-established homeowner associations as the best way to ensure that ever-larger and more elaborately packaged subdivisions and residential complexes will maintain their character until build-out and beyond. The Urban Land Institute's *Residential Development Handbook* warns developers that "in many ways, . . . governance is one of the most important parts of project planning and execution."[15] Initially concerned chiefly with the preservation of the aesthetics and overall design vision of high end developments, the new generation of common-interest communities soon moved to defend their residential niches against unwanted development (such as industry, apartments, and offices) and then, as environmental quality became an increasingly important social value, against any kind of development. This "Sunbelt Bolshevism," as Davis called it in the context of Southern California, became an important element in the no-growth/slow-growth politics of American suburbs: "the latest incarnation of a middle-class political subjectivity that fitfully constitutes and reconstitutes itself every few years around the defense of household equity and residential privilege."[16] At the same time, homeowner associations established themselves as regular participants at public meetings of city councils, school districts, and planning boards. Complaining about encroachment and undesirable development, they represent the vanguard of NIMBYism.

It follows that homeowner associations can also be an important element in regime theory, which attempts to examine how various coalitions come together to achieve particular short- or medium-term outcomes.[17] Frequently these are the interests of growth machine coalitions put together by developers and political entrepreneurs. The crucial point about regime theory is that power does not flow automatically but has to be actively acquired. In the context of economic restructuring and metropolitan change, city officials seek alliances, it is argued, that will enhance their ability to achieve visible policy results. These alliances between public officials and private actors constitute regimes through which governance rests less on formal authority than on loosely structured arrangements and dealmaking. With an intensification of economic and social change in metroburbia, new sociopolitical cleavages—green, yuppie, populist, neoliberal—have been added to traditional income-, class-, and race-based cleavages, so that regimes have become more complex and, potentially, more volatile. Meanwhile, the scale and extent of economic restructuring has meant that greater competition among municipalities for economic development investments has added a new dynamic, whereby the intensity of political conflict within them is muted.

Homeowner Associations and Local Democracy.   Robert Nelson characterizes the privatization of neighborhood governance through homeowner associations as "a new exercise in constitution writing on a scale larger than this country has ever seen before." He compares the significance of this "transformation of local government" to the transformation of business more than a century ago, when the rise of the modern corporation restructured the relationship between the ownership of business property and the managerial control of business. Writing from an economic and legalistic perspective, Nelson wrestles for 450 pages with the theoretical niceties surrounding individual and property rights within an increasingly libertarian and neoliberal environment. He points out that, in theory, "a world of free markets . . . is a world in which real neighborhood autonomy can have little place," but concludes that private neighborhood associations "will become a central political institution in America in the twentyfirst century."[18] McKenzie emphasizes that "this trend is not a passing fashion but an institutional transformation reflecting the ideological shift toward privatism characteristic of the neoliberal consensus."[19]

Nelson—who ends his book with a call for the pursuit of neoliberal urban policies like public service pricing, educational competition, freedom of local government organization, private delivery of public services, and the deregulation of subdivision controls—concludes that the implications of the trend are broadly beneficent. For suburban municipalities, common-interest communities are generally a good thing because homeowner associations typically arrange for trash collection, remove leaves in the fall, plow snow in the winter, light streets, tend parks and gardens, and provide other services and amenities that would otherwise fall to public authorities. "Thus cities can acquire new property taxpayers without having to extend to them the full panoply of municipal services."[20]

But others see the rise of private neighborhood associations as a cause for concern. Alex Marshall suggests that local government is being traded for corporate management. Ed Blakely and Mary Gail Snyder, echoing Robert Reich, describe the trend as "civic secession." Evan McKenzie argues that "the rise of residential private government facilitates the emergence of a two-tier society in which the 'haves' are increasingly separated—spatially, institutionally, socially and economically—from those of lesser means. I call this realm 'privatopia' because it represents the pursuit of utopian aspirations through privatization of public life. Within privatopia the terms and conditions of life are at odds with the norms and expectations of liberal democracy."[21]

Servitude Regimes: No More Pink Flamingos. Servitude regimes—the set of covenants, controls, and restrictions attached to the deeds of property in a common-interest community—are often cited as being at odds with American ideals of individual freedom. As McKenzie put it: "No more pink flamingos." Servitude regimes are typically designed by developers not only to preserve landscaping and maintain the integrity of urban design but also to control the details of residents' homes and their personal comportment. Developers thus become benevolent dictators, imposing a bourgeois cultural framework on the landscapes and communities of the New Metropolis. For consumers, these servitude regimes offer a means of narrowing uncertainty, protecting equity values, and, above all, establishing the physical framework for the material consumption that constitutes their lifestyle. Little is left to chance, with covenants, controls, and restrictions detailing what is and what is not allowed in terms of garden fences, decks, hot tubs, and clotheslines, the color of doors and mailboxes, and so on. Most ban all signs except for real estate placards and restrict what kind of vehicles can be parked outside, even in driveways; some even prescribe how long garage doors can be left open, the type of furniture that can be seen through front windows, the color of Christmas tree lights, and the maximum length of stay for guests. Most limit the number and types of pets that residents may keep, as well as the kinds of activities that are allowed in gardens, driveways, streets, and public spaces, and whether any sort of business can be conducted from the home.

As noted in chapter 4, covenants, controls, and restrictions have to be taken very seriously under U.S. law, since if homeowner associations don't enforce them to the letter they can be accused of being arbitrary and capricious. Homeowner associations have the power to levy fines in order to bring transgressors into compliance. They also have the power to obtain a lien on the property of recalcitrant homeowners, and can even threaten foreclosure. Because of the all-or-nothing nature of servitude regimes under the law, stories of draconian responses to petty misdemeanors have become commonplace in local newspapers and law journals.

- In Boca Raton, Florida, a couple was ordered to stop using their backdoor because they were wearing down a path in the grass on their approach.[22]
- In Ashland, Massachusetts, a Vietnam War veteran was told he couldn't fly the American flag on Flag Day.[23]

- In Montgomery County, Maryland, Rick McCann and his wife, Julia, erected a wooden fence around their house in Potomac to keep their young children out of traffic. This ran afoul of their homeowner association rules requiring that fences be approved by a committee of neighbors. The association sued the McCanns, asking a judge to force its removal. After $25,000 in legal fees, the couple retained their fence, though a portion had to be moved as part of an out-of-court settlement.[24]
- Elsewhere in the same county, a townhouse owner, ordered by homeowner association board members to stop parking his car in a visitor's lot, responded by moving it to a fire lane and then onto his front lawn—both proscribed by the development's covenants, controls, and restrictions. The feud escalated when he put up an unauthorized basketball hoop. An association leader, complaining at one point that he was being stalked, got a court order barring the homeowner from any contact, including at the school that both their children attended.[25]
- In Calaveras County, California, a retired couple failed to pay $120 in association fees due in January 2003; in December the same year, their homeowner association in Copperopolis foreclosed and sold their home.[26]
- In New Jersey, Margaret and Haim Bar-Akiva were fined by their Twin Rivers Homeowners Association in 1993 because the front door on a house they had bought for Ms. Bar-Akiva's parents was painted the wrong color. They were then fined because the bars on their storm door violated association guidelines. In the ensuing dispute over the storm door, the association spent $100,000 of the community's money on lawyers—who kept the couple in depositions for seven hours. Then, in 2000, the Bar-Akivas were ordered to remove campaign signs they had put on their lawn in the run-up to a homeowner board election. Mr. Bar-Akiva claimed they were singled out because the signs supported a dissident slate; the board countered that signs were permitted only in windows and garden beds bordering each house.[27] The case went all the way to the New Jersey Supreme Court, which ruled in 2007 in favor of the homeowner association.

As Rob Lang and Jennifer LeFurgy observe, "a few HOAs got carried away in attempting to regulate *anything* that could possibly diminish home values. They also became a bit trigger happy and ready to foreclose on residents for the failure to pay fines on tiny infractions."[28] A survey conducted in California in 2001 by Oakland-based Sentinel Fair Housing examined foreclosures in Alameda, Contra Costa, San Mateo, Santa Clara, and Sacramento counties. It found that the median amount of back assessments due on foreclosures conducted in private communities was $2,557. In contrast, the subjects of more than 4,000 foreclosures outside such communities owed a median of $190,000.[29] As a result of this kind of zealousness, some states have begun to set boundaries on homeowner associations. In Nevada, an ombudsman program for homeowner associations handles 3,000 complaints annually. In California, legislation was proposed that would allow homeowners to run up as much as $50,000 in unpaid dues and fines before a homeowner association could take foreclosure action. The proposal was defeated, but in 2005 California created an ombudsman position to manage disputes between homeowner associations and residents and passed a bill that gives homeowners the right to redeem their properties for ninety days after a nonjudicial foreclosure and requires a minimum amount of $1,800 in unpaid assessments. In Arizona, homeowner associations must give three years' grace to delinquent homeowners and then secure a court order prior to placing a lien on a home. In addition, homeowners have their disputes heard before an administrative law judge rather than having to resort to more expensive and time-consuming litigation in Superior Court.

Homeowner Associations, Citizenship, and Democracy.   The governance of homeowner associations has raised a number of issues, both in principle and in practice. The boards of homeowner associations are directly elected by property owners, though it is on the basis of one homeowner vote per unit of property, rather than per adult, which calls into question the nature of "citizenship" in master-planned communities. Another issue of representation arises in large developments where a high percentage of one type of housing may result in the residents of that type of housing dominating the governance process because they have sufficient votes to control the election of the board. Furthermore, it is standard for developers to get three votes in a homeowner association for every unit of property that they own. This effectively allows the developer to control the association until three-quarters of all the property is sold.

Homeowner boards can modify the covenants, controls, and restrictions, add new ones or delete existing ones, but such action typically requires a supermajority of all property owners (not just those who choose to vote). In practice, homeowner associations generally have very low turnouts—at best, comparable to participation rates in local school board elections—and many have a scattering of absentee owners, so that substantive change is unlikely and unusual. Servitude regimes, in other words, are effectively rigid, with little prospect for association boards to develop a legislative program of any significance. The result is that most residents behave like stockholders in a large corporation, not knowing or caring much about the details of the operation as long as their stock (the value of their property, in this case) remains strong.

Boards, meanwhile, are left to ensure the maintenance of property values by enforcing the covenants, controls, and restrictions. In practice, identifying transgressions of the covenants, controls, and restrictions is usually left to volunteers, who are often appointed at poorly attended meetings and who may or may not have any legal or governmental experience. Some larger and more affluent developments hire people to cruise around in order to spot violations. In addition, of course, some infractions are called in by irate or indignant neighbors. While some disputes can escalate into the legal equivalent of road rage, most cases that come before homeowner boards amount to genteel struggles over parking restrictions, signage, commercial vehicles, the use of public spaces, paint colors, satellite dishes, basketball hoops, mailboxes, play equipment, and flag flying. The common denominators are equity values and implied norms about whether or not the development looks nice, neat, and trim; and implicitly whether the residents are tasteful and refined. Yet the scope of their activities permits the regulation of a much wider range of behavior than any public local government. Who would countenance, for example, any public agency presiding over the details of interior furnishings, house guests, pets, and garden furniture? Not to mention overt discrimination by age (e.g., not allowing younger households to live in "active adult" communities), which would be plainly illegal in the public sector.

Equally unnerving from the point of view of democratic process is that in enforcing their covenants, controls, and restrictions homeowner boards are able to act as accuser, judge, and jury. Though their day-to-day issues may seem petty, homeowner associations operate in ways that many consider worryingly undemocratic. McKenzie argues that "boards of directors operate outside constitutional restrictions because the law views them as business entities rather than governments."[30] Defenders of private neighborhood government argue that people consciously consent to covenants, controls, and restrictions by buying property and living in a common-interest community.[31] McKenzie and others regard this as a "legal fiction," pointing out that consent is not simply a legal issue of someone having signed on the dotted line (regardless of whether or not they have read through dozens of pages of covenants and restrictions) but, rather, a matter of having little choice. So much of the housing in the middle- and upper-middle price range is subject to covenants, controls, and restrictions that households have little real choice. Some municipalities—boomburbs like Coral Springs, Florida, Henderson and North Las Vegas, Nevada, and Chandler and Gilbert, Arizona—allow new development only in the form of common-interest communities. Others are trying to retrofit a common-interest structure over existing neighborhoods with fee-simple land ownership.[32] Critics of private neighborhood government also point out that in their servitude regimes the concept of individual rights is replaced with that of restrictions, and that the notion of civil society, implying responsibilities to the broader community, is reduced to nothing more than meeting one's economic obligations and conforming to rules that are framed around the protection of property values. There is a broader issue, too, in terms of the accountability and obligations of private neighborhood governments to public governmental agencies. In particular, their lack of accountability has become a serious impediment to regional and metropolitan-wide planning (ironically, regional-scale planning is one of the big planks of the new urbanist platform). Cellular accretions of master-planned communities, many with private roads, walls, and gates, defy any aspiration toward strategic land-use planning and make it effectively impossible to operationalize metropolitan transportation systems or even to build new roads.

All in all, it's quite a phenomenon: a country that prides itself on the primacy of individual rights and that sees itself as the avatar of democracy has quietly forfeited its dearest principles in the cause of equity values and the sumptuary codes of the affluent middle classes. What makes this even more intriguing is that there has been a simultaneous resurgence of libertarianism, especially among the class fractions that populate private master-planned communities.

Libertarianism, Property Rights, and Zoning

The 1970s and 1980s saw the rise of a strong libertarian element in American political and economic thought. Individual liberty and the sanctity of private property ownership were the cornerstones of key theoretical expositions that sought to undermine the rationale for land-use zoning.[33] In the neoliberal political climate of the 1980s, popular support for progressive notions of scientific management had ebbed away, to be replaced by a greater emphasis on individual self-expression and a new resistance to central authority. In populist politics as well as in law and economics journals, the traditional arguments for land-use zoning—abating nuisances, creating efficiencies in planning for the common good—were increasingly dismissed. Instead, zoning was portrayed as an inequitable redistribution of the rights to develop land. The argument began to form among libertarian theorists that municipal zoning amounts to a "taking" of property owners' development rights that should not be allowed, or that should be compensated financially.

Resolving this sort of thinking with the case for private neighborhood governments in which individual freedoms and property rights are constrained by rigid covenants, controls, and restrictions requires a certain amount of mental gymnastics. Here it goes: residents of private neighborhood governments are exercising their individual freedom by choosing to live in a local regime of restricted freedoms, in order to gain collective control over the actions and behaviors of their neighbors. That's OK then.

Meanwhile, buoyed by the political groundswell of neoliberalism, libertarians and property-rights activists have attempted to use Congress, state legislatures, and ballot initiatives to pass laws that would treat zoning and other land-use regulations as "takings," requiring financial compensation to affected property owners in the same way that property owners affected by planners' use of powers of eminent domain are compensated. Their first significant win came in November 2004, when Oregon voters passed Measure 37, a ballot initiative that torpedoed what had been the most sophisticated land-use planning system in the country. Under Measure 37, zoning decisions as well as the use of eminent domain are regarded as "regulatory takings," and affected property owners must be compensated for the loss of their notional development rights. The following year, however, a Supreme Court ruling effectively broadened the use of eminent domain to cover economic development in addition to its traditional uses in projects that benefit public health, safety, and environmental protection. The outcome of this case, *Kelo v. New London*, held that the town of New London, Connecticut, could condemn the homes of Susette Kelo and six other holdouts in order to make room for the redevelopment of the area surrounding a global pharmaceutical company's hundred-acre manufacturing complex with a hotel/conference center, industrial park, and residences designed to complement the manufacturing complex. To many, this extended government powers too far, and it encouraged libertarians to launch ballot initiatives with titles like "Protect Our Homes," "The Home Owners' Protection Effort," and "People's Initiative to Stop the Taking of Our Land" in order to advance a broader anti-government agenda. In the November 2006 elections, property rights initiatives were passed in Arizona, Florida, Georgia, Michigan, Nevada, North Dakota, Oregon, and South Carolina, in most cases by an overwhelming margin. These successful initiatives fell short, however, of the aspirations of libertarian groups like the Reason Foundation, Americans for Limited Government, and the Cato Institute, which would like to see more initiatives like Oregon's Measure 37.

The implications of libertarian takings initiatives for policy and planning—and for the evolution of the landscapes of the New Metropolis—are enormous. Unable to find the resources with which to compensate private property owners for regulatory takings, municipalities would be compelled to allow myriad exceptions to their regulatory frameworks, thus replacing any semblance of strategic planning with piecemeal negotiations over every land-use decision. Municipal planners would be emasculated; consultants would rule (and collect handsome fees).

In private master-planned developments, meanwhile, the takings issue will be irrelevant: the servitude regimes of covenants, controls, and restrictions insulate and exempt them from most municipal land-use planning regulations, making them even more attractive to developers. Future iterations of American suburban landscape development are therefore likely to be framed overwhelmingly within private master-planned developments. We are thus confronted with a breathtaking only-in-America irony: More and more people living and working in communities with a significant circumscription of individual rights, thanks to the success of the libertarian movement.

The Geopolitics of Suburbia

Even though libertarians may not be happy at the continued existence of local municipalities' powers of zoning and eminent domain, greenfield sites in most of suburbia are subject to almost no strategic planning restrictions (such as growth boundaries); most land is zoned at densities at which residential development can be profitable; and it is usually possible to get a zoning variance for the kind of retail, personal service, and office functions that can be profitable in suburban settings. The problem for both property owners and developers arises because of the hop-scotch nature of sprawl, jumping out to subdivide land amid rural communities, or filling in patches of undeveloped land adjacent to established subdivisions and master-planned communities. Whenever new development is proposed it is likely that someone will feel that their property values and/or lifestyles are threatened. This being America, it is also likely that the lawsuits will follow. Bill Fischel, an economist at Dartmouth College, has developed the concept of the "homevoter."[34] Fischel argues that homeowners are especially conservative because what is typically their single largest asset—their home—is always linked to the fate of their neighborhood. The result is that they tend always to vote in local politics in the interest of resisting change that might affect their neighborhood.

Meanwhile, metropolitan restructuring and the neoliberal distaste for taxation and public spending has left many local municipalities experiencing fiscal stress. In their weakened position, most municipalities have privatized many of the functions and responsibilities that they had taken on during the expansion of egalitarian liberalism. In the vacuum left by this retreat, voluntarism has become a principal means of providing for the needs of the indigent, while in more affluent communities homeowner associations have taken on the burden of service and amenity provision. Public local governments have turned increasingly to the private sector for capital for economic and social investment through public-private partnerships of various kinds, and now give much greater priority to economic development than to the traditional service-providing and regulating functions of the local state. This civic entrepreneurialism has fostered a speculative and piecemeal approach to the management of metropolises. In central cities the emphasis has been on set-piece projects such as downtown shopping centers, festival market places, conference and exhibition centers and the like, that are seen as having the greatest capacity to enhance property values (and so bolster the local tax base) and generate retail turnover and employment growth. In suburban jurisdictions, the emphasis has been on attracting upscale office and retail employment while resisting any influx of residents with high levels of need and low levels of tax capacity.

The decentralization of jobs and residences has also brought about a corresponding decentralization and proliferation of local jurisdictions. This process has been accelerated by federal and state policies that, guided by the principle of local autonomy, make the annexation of territory by existing cities more difficult while keeping incorporation procedures very easy. The fragmentation of general-purpose government in the United States has also led to the suppression of political conflict between social groups. As political scientist Ken Newton has observed, social groups can confront each other when they are in the same political arena, but this possibility is reduced when they are separated into different jurisdictions. "Political differences are easier to express when groups occupy the same political system and share the same political institutions, but this is more difficult when the groups are divided by political boundaries and do not contest the same elections, do not fight for control of the same elected offices, do not contest public polities for the same political units, or do not argue about the same municipal budgets."[35] This attenuation of democracy means in turn that community politics tends to be low key, while the politics of the whole metropolitan area are often notable for their absence. The balkanization of the city means that it is difficult to make, or even think about, area-wide decisions for area-wide problems. The result is that small issues rule the day for want of a political struc-ture that could handle anything larger.

NIMBY Politics

NIMBY ("Not In My Backyard") squabbles are a chronic aspect of local politics throughout suburbia. When changes to the physical or social fabric of a neighborhood are perceived by more than a handful of residents to threaten property values—or, more euphemistically, the "character" of a neighborhood—the flashpoints for NIMBY squabbles become politicized. Homevoters, faced with the prospect of some unwanted change, are typically quick off the mark with placards, leaflets, and mobilization emails with upper-case phrases and multiple exclamation points for emphasis. At issue, almost always, are property values and the closely associated concerns of aesthetics and social exclusion.

The appetite for bigness and bling that has led to the insertion of big new homes—"McMansions"—in place of older homes in established neighbor-hoods, for example, has led not only to NIMBY bickering among neighbors but also to political campaigns aimed at changes in zoning ordinances. Thus, for example, Chevy Chase, Maryland, an upscale suburb of Washington, D.C., imposed a six-month moratorium on home construction in 2005 to make time to examine how to deal with the proliferation of oversized single-family houses on "scrapeoff" sites. In 2006, the Los Angeles city council passed an ordinance that limits home size on lots of 8,000 square feet or less in the Sunlund-Tujunga area to 2,400 square feet or a floor-to-area ratio (FAR—really the house footprint-to-lot ratio) of 40 percent, whichever is greater. New Canaan, Connecticut, enacted regulations in 2005 that limit the height of new houses, while Austin, Texas, has introduced floor-to-area ratio limits of 40 percent on new housing, a maximum of 2,500 square feet for replacement construction, and a maximum of 20 percent of existing floor space for additions.

Social and demographic change produces similar reactions. In Manassas, Virginia, thirty-five miles southwest of downtown Washington, D.C., the cost of housing has meant that some extended families decided to share space in single-family homes. Many of these families included immigrants working in Manassas and unable to find affordable housing any other way within reasonable commuting distance. This was entirely legal until neighbors began to complain to city hall in Manassas. The city's rather hysterical response was to set up an "overcrowding hotline," while the mayor sent letters to the governor of Virginia asking him to declare a state of emergency. When this failed, the city of Manassas decided to redefine its definition of a "family," adopting a zoning ordi-nance in December 2005 that essentially restricts households to immediate rela-tives, even when the total number of persons in the household is below the city's occupancy limit.

Exurbia and the rural-urban fringe, meanwhile, has become a land-use battleground, where "developers, long-term landowners, quick-buck land spec-ulators, politicians and realtors are matched against other long-term landowners, politicians, environmentalists, and newcomers who want to keep their communities attractive and fiscally manageable."[36] Dolores Hayden has noted the paradox that successful NIMBY campaigns against new developments can sometimes accelerate suburban sprawl, pushing unwanted development farther out into greenfields.[37] This, of course, is another reason for a regional approach to planning and development.

A little more predictability and coordination would also benefit developers. Toll Brothers, the Pennsylvania-based luxury home developer, has estimated that 90 percent of the company's recent projects in New Jersey have involved some sort of legal entanglement. Obtaining approvals and negotiating community objections can take three to five years in some states, and even in Florida and Virginia—traditionally easy places to build—clearing the approval process can take a year or two. One Toll Brothers development in New Jersey—The Estates at Princeton Junction—took ten years from the time the land was acquired until the company was allowed to break ground. Even before a single bulldozer arrived to clear topsoil, the company's legal bill had amounted to $2 million. The neighboring township of West Windsor had balked at Toll's plan for the development, so Toll sued, invoking a New Jersey law whereby developers can push a project forward if its plan includes affordable housing in a locality where such a need exists. Toll won the case in 1997 but it took until 2002 before the project could go ahead because appeals went all the way to the state's Supreme Court.[38]

Zoning Struggles.   Much of the political struggle in suburbia turns on how to manipulate local zoning laws in order to protect or achieve some sort of advantage. For affluent neighborhoods in areas designated for low-density residential development, zoning acts as an invisible wall, keeping out undesirable households and land uses. Only those who already own property in the municipality have standing in court to bring suit against the zoning codes, so outsiders are unable to effect change. As a result, as James and Nancy Duncan noted in their study of Bedford, New York, zoning "plays an active structuring role in grounding the practice of an aestheticized way of life in a place. It attempts to maintain sufficient social homogeneity within a territorially bounded and (relatively) defensible space in order to achieve a collective sense of place and landscape."[39] At the same time, low-density zoning pushes up house prices, "zoning out" key workers with modest salaries—teachers, nurses, emergency services workers, and so on—and exacerbating metropolitan transportation problems.[40]

From the perspective of local municipalities, exclusionary zoning is an important tool in competing with other jurisdictions for fiscal health. The goal is to use zoning, along with other land-use planning tools, to keep out burdensome populations (with high levels of needs but low tax capacity) and noxious industries while attracting affluent and self-sufficient populations and clean economic activities that pay their workforce well. The recipe for success in this fiscal mercantilism is a moderate amount of upscale office park development and perhaps some high-end retail development, rounded out by master-planned developments with homeowner associations that take care of as many local services as possible. But there is only so much high-end office and retail development to go around, and even with developers paying for streets and sidewalks (and, sometimes, school buildings) and homeowner associations footing the bill for trash collection, snow clearing, landscaping, and so on, the revenues from property taxes may not be sufficient to cover the costs of installing roads, water, and sewer hookups for low-density sprawl, or of maintaining schools, police and fire departments, and other public services.

In the end, of course, municipal policies are decided by elected boards, so that the need to balance the books through carefully targeted growth is set against voters'—mostly "homevoters" in suburban jurisdictions—preference for low taxes and their sensitivity to their property values. The result is a struggle between growth-machine coalitions on one side and slow-growth or no-growth coalitions on the other. Loudoun County, Virginia, provides a classic example. Rapid growth during the 1990s increased the county's population from 86,000 to 170,000: a boom time for local builders, developers, and other members of the local Chamber of Commerce but alarming for both residents and incomers who were witnessing a dramatic escalation of traffic congestion and increasing signs of stress in the county's ability to provide the kind of schooling and public services that they had come to expect. Led by a grassroots citizens group, Voters to Stop Sprawl, voters installed a slow-growth set of representatives in the 1999 elections. The new leadership promptly rezoned large swaths of undeveloped land, effectively placing the western two-thirds of the county off limits to conventional subdivisions, with developers generally limited to no more than one home per twenty acres in northwest Loudoun County (or one house per ten acres in cluster subdivisions with lots of surrounding green space). Pro-growth activists challenged the rezoning in Virginia's Supreme Court, which eventually ruled in their favor on a technicality: that the county had not clearly defined in its public notices the boundaries of land to be rezoned. Meanwhile, a new crop of pro-growth officials took power in Loudoun after elections in 2004, underwritten by campaign contributions from a growth-machine coalition. One of the first rules the new county board adopted was to take away the agenda-setting function from its chairman, a hold-over slow-growth advocate, and place it into the hands of the vice-chairman, a pro-growther.

"Smart" Growth.   An alternative to the continual and unappetizing politics of this sort has emerged as a sort of "third way" politics: have your cake and eat it too. It is the doctrine of "smart growth," often attributed to Robert Yaro, now president of the Regional Plan Association, when he worked as a planner in the administration of Massachusetts governor Michael Dukakis in the 1970s. Smart growth is pro-growth, but only when it is relatively compact and steered toward strategically designated locales with adequate infrastructure. It is, in other words, a stealthy euphemism for old-fashioned regional planning and growth management of the sort that cannot be entertained in the lexicon of a

neoliberal political economy. Smart growth adheres to the principles of preserving public goods, minimizing adverse land-use impacts, maximizing positive land use impacts, minimizing public fiscal costs, and maximizing social equity[41]—hard to disagree with, then. It has been endorsed by the U.S. Environmental Protection Agency (EPA), the Lincoln Institute of Land Policy, and the National Resources Defense Council and has been gratefully embraced by an embattled planning profession that has been under the heel of neoliberalism. It has also been endorsed by Andres Duany and his new urbanists, who support it because at face value it reinforces their platform.

A national coalition, Smart Growth America, has attracted dozens of member organizations, including the American Farmland Trust, the American Planning Association, the Congress for The New Urbanism, 1000 Friends of Connecticut, 10,000 (not to be outdone) Friends of Pennsylvania, the Sierra Club, and the Trust for Public Land.[42] A Smart Growth Network,[43] coordinated by the Environmental Protection Agency's Division of Development, Community and Environment, in conjunction with several nonprofit and government organizations, including the International City/County Management Association and the Sustainable Communities Network, has developed a set of ten basic goals (referred to as "principles") for smart growth: mixed land uses; taking advantage of compact (i.e., higher-density) neighborhood design; creating housing opportunities and choices, creating walkable communities; fostering distinctive communities with a strong sense of place; preserving open space, farmland, and critical environmental areas; strengthening and directing development toward existing communities; providing a variety of transportation choices; making development decisions predictable, fair, and cost-effective; and encouraging community and stakeholder collaboration in developer decisions. It is blithely asserted that elements of this wish list "can be applied in various combinations to create smart, nonsprawling communities."[44]

Most prominently, Parris Glendening, governor of Maryland between 1995 and 2003, made smart growth the center of his electoral platform. In office, he appointed a cabinet secretary to oversee development policy, pulling together the state agencies—transportation, housing, environmental quality—that had anything to do with growth. Glendening insisted that the state itself take the lead in smart-growth-type policies, locating state agency offices only in downtowns and town centers. He redirected state funding from highways to transit and to infrastructure in higher-density settings, and championed a policy under which developers paid for water, sewer, and other infrastructure in undeveloped areas, while developers building in designated higher-density areas benefited from streamlined approval processes and reduced fees.

Naturally, policies like these were a rude and unwanted provocation to progrowth interests. Libertarian and neoliberal think-tanks like the Heritage Foundation and the Reason Foundation cranked out essays warning of the anti-American implications of smart-growth "abuses"[45] (sometimes portrayed as elitist, sometimes as socialistic), damage to free-market mechanisms, and constraints on individual choice; they provided lobbyists with lists of worrisome key points to use in speaking to legislators, and fed talking points to op-ed writers in local newspapers. Developers simply stepped up their campaign contributions to pro-growth candidates in local elections. But, to the dismay of smart-growthers, the most effective challenges to their policies came from citizens themselves, in classic NIMBY responses. In Maryland, for example, residential and retail projects around Metro stations—considered ideal locations for smart growth because they would encourage the use of mass transit—have been stopped cold or scaled back because of neighborhood opposition in Takoma Park and Tenleytown. Even projects in designated residential smart-growth areas have run into local opposition. Maple Lawn Farms, a 508-acre site in Howard County three miles south of Columbia, Maryland, midway between the converging metropolitan areas of Baltimore and Washington, and with a six-lane highway running along one side and subdivisions wrapped around the other three sides, is a classic smart-growth site. But neighbors—including the former chairman of the Howard County chapter of the Sierra Club!—objected, contending that they already faced crowded roads and schools and needed to preserve the remaining open space in the area.[46] After thirty-two public hearings, construction began in Maple Lawn Farms in 2004. Instead of the initial proposed density of three homes per acre—fairly typical for American suburbs but on the low side for a smart-growth project—the final approved density was 2.2 homes per acre. By the time Glendening left office in 2003, his office had concluded that the rate at which farm and forest land was being developed in Maryland had not slowed appreciably.

Who Pays, Wins

Under the radar of formal politics, though, are nods and handshakes that can be just as important in shaping outcomes. Pro-growth coalitions have consistently been able to keep housing and real estate markets biased in favor of sprawl, mostly through business linkages and networks. In particular, many elected officials have links to the building and development industry through seats on bank boards, stock ownership, and campaign contributions. Most of these relationships are not illegal, though to outsiders they may seem like obvious conflicts of interest; and there are certainly plenty of cases of influence peddling, sharp practice, and corruption that support a skeptical view of the relationships between local politics and the development industry.

Take, for example, the pro-growth coalition in Loudoun County, Virginia. A series of investigative articles in the *Washington Post* in 2007 detailed how major land-use decisions in Loudoun have been dominated by a small network of public officials and their allies in the development industry.[47] By coordinating with developers, landowners, and other real estate interests, pro-growth officials who took power in 2004 have served as gatekeepers to a wave of suburban development, sometimes following detailed instructions from those in the development industry on how to vote and what to say in public. According to the *Post*'s investigation, members of the new pro-growth board voted soon after they took office to boost the number of homes that could be built on the family farm of Dale P. Myers, a former supervisor who had been instrumental in getting many of them elected. A month later, a builder bought the property from Myers's family for $12.2 million—four times the assessed value of the property before the zoning decision. Next, the board agreed to authorize the county to purchase a different parcel for $13.5 million, once again helping Myers, who was acting this time as the real estate agent.

As chairperson of Loudoun's Board of Supervisors in the late 1990s, Myers had been a persistent advocate for speeding up development. After losing her place on the Board in the 1999 slow-growth election sweep, she started a consulting firm and also became a key organizer of Citizens for Property Rights, a lobbying group that targeted county supervisors who supported building limits, sponsoring attack ads and sharply criticizing them at public meetings. Meanwhile, Loudoun's slow-growth board had passed strict development limits on large tracts of the western reaches of the county. In response, developer Greenvest L. C. (whose companies owned more land in Loudoun than any other private landowner—more than 5,000 acres, and much of it undeveloped) filed more than twenty lawsuits against the county, part of a coordinated barrage of about two hundred such suits by developers seeking to overturn zoning restrictions.

For her part, Myers, along with others in the real estate development industry, took key positions in Loudoun's Republican Party and backed candidates who would provide reliable votes for development. Myers became a top strategist for three candidates: Bruce E. Tulloch, Stephen J. Snow, and Lawrence S. Beerman II, a former county supervisor. Counseled by Myers, Greenvest supported their candidacy. According to the *Washington Post* investigative team, Greenvest chief executive Jim Duszynski said that the goal was "regime change." In the run-up to the 2003 election, Greenvest companies and executives contributed $18,000 to the campaigns of six pro-growth board candidates, along with indirect contributions to the cause through state and local Republican organizations. Overall, according to public records obtained by the *Post*, companies and individuals tied to the development industry poured more than $490,000 into 2003 supervisor campaigns in Loudoun, more than seven times the figure four years earlier. Duly elected, the new supervisors met for the first time in January 2004 and took an initial vote to allow central sewer and water lines throughout a 23,000-acre patch of western Loudoun County that had been downzoned by the slow-growth board. It was a key step in an effort to open up a vast area to builders; Greenvest had purchased thousands of acres there. In its first two years, the new pro-growth board approved more than 9,000 new homes. In four key votes, Greenvest gained permission to build more than 1,800 homes in addition to the roughly 300 that would have been allowed under the zoning that had been approved by the slow-growth board.

CHAPTER 7

Material Culture and Society
in Metroburbia

To really understand America, as the conservative commentator David Brooks observes, "You have to take seriously that central cliché of American life: the American Dream."[1] The entire residential fabric of metroburbia rests on the American Dream, founded on the promise of ever-increasing levels of material consumption, the expectation of single-family home ownership, the accumulation of wealth, and systematic upward social mobility through ingenuity and hard work. Brooks is particularly taken with the way in which material things "are shot through with enchantment":

> The suburbs themselves were built as conservative utopias. Children are raised with visions of ideal lives. This is the nation of Hollywood, Las Vegas, professional wrestling, and Disney, not to mention all the other fantasy factories. This is the land of . . . the whole range of ampersand magazines (*Town & Country, Travel & Leisure, Food & Wine*) that display perfect parties, perfect vacations, and perfect meals—ways of living that couldn't possibly exist in real life. This is the land of Rainforest Café theme restaurants, comic-book superheroes, Shangri-La resort hotels, Ralph Lauren WASP-fantasy fashions, Civil War reenactors, gated communities with names like Sherwood Forest Grove, and vehicles with names like Yukon, Durango, Expedition, and Mustang, as if their accountant-owners were going to chase down some cattle rustlers on the way to the Piggly Wiggly.[2]

Materialism and Social Change

For Brooks, this is all symptomatic of a "Paradise Spell," an imaginative drive that lies behind Americans' energetic attempts to live out the Dream and the reason for their tendency to work so hard and consume so feverishly. In academic terminology, it is the spell of "self-illusory hedonism" identified by sociologist Colin Campbell (see chapter 1).[3] In a more empirical vein, economic historians point to the 1920s as the moment when consumers' aspirations and purchasing power began to converge, as Fordism unleashed a new sociocultural phenomenon: competitive consumption. This was also the moment when realtors' Own-Your-Own-Home campaign solidified the idea of the home as a privileged consumer durable, the cornerstone of the original version of the American Dream, the stage for materialistic lifestyles and the container for an extended range of material possessions. In the economic boom after World War II, the Dream took on a more expansive form as discretionary spending by the middle classes reached unprecedented levels. Harvard economist James Duesenberry identified the trend at an early stage, contrasting it with the nineteenth-century version of conspicuous consumption that had been documented by Thorstein Veblen. Instead of being driven by an elite "leisure class," post–World War II consumption was a middle-class suburban phenomenon, driven by neighbors: the eponymous "Joneses."[4]

By the 1960s, Guy Debord had identified the emergence in Western culture of a *Society of the Spectacle*, defined as the "moment when the commodity has attained the total occupation of social life."[5] Jean Baudrillard wrote of "the need to need, the desire to desire."[6] Nowhere was this more evident than in the United States, where baby boomers were coming of age and transforming American norms of consumption as well as politics and popular culture. The formative experience of the baby boomers was the postwar economic boom. Growing up in affluent sitcom suburbs, they initially rebelled against the apparent complacency of what John Kenneth Galbraith dubbed the "Affluent Society," channeling their energies into counter-cultural movements, many of them with a vaguely collectivist approach to the exploration of freedom and self-realization. The high-water mark of this "alienation generation" was 1968, the year of sit-ins, protest marches, strikes, civil disorder, and riots. The heady mix did not last long. By 1973, the global economic system-shock triggered by the quadrupling of crude oil prices by the Organization of the Petroleum Exporting Countries had sobered the boomers into a more self-oriented perspective. The failure of 1960s radicalism, observes Agnes Heller, produced the preconditions for the emergence of a "postmodern generation" characterized not by collectivist idealism but by self-oriented materialism.[7] David Ley characterized the turnaround as follows:

> The post-modern project is the project of the new cultural class, representatives of the arts and the soft professions who came to political awareness in the 1960s and were receptive to the oppositional ideas of the counter-culture. But through the 1970s and 1980s hippies have all too readily become yuppies, as the subjective philosophies of phenomenology and existentialism which opposed the impersonality of modernism in the 1960s and redeemed the individual have been directed inwards, and the celebration of meaning has shifted subtly to the celebration of meaning of the self.[8]

Very quickly, self-awareness became a commodity as well as a state of mind. New product lines, promoted by magazines like *Self*, emerged to cater to the trend; and advertising strategies began to exploit individual preferences and desires. In the mid-1970s maturing baby boomers found themselves flooding labor and housing markets just as the economy was experiencing the worst recession since the 1930s. Salaries stood still while housing prices ballooned. In 1973, "the last really good year for the middle class, the average 30-year-old man could meet the mortgage payments on a median-priced house with about a fifth of his income. By 1986, the same home took twice as much of his income. In the same years, the real median income of all families headed by someone under 30 fell by 26 percent." It prompted the apostasy of the postmodern generation: "In our 20s, my friends and I hardly cared. We finished college (paid for primarily by our parents), ate tofu, and hung Indian bedspreads in rented apartments. We were young; it was a lark. We scorned consumerism. But in our 30s, as we married or got sick of having apartments sold out from under us, we wanted nice things, we wanted houses."[9]

Yet economic circumstances did not often permit a smooth transition to materialism, even for the college-educated middle classes: "My friends dressed and ate well, but despite our expensive educations, most had only one or two elements of the Dream we had all laughed at in our 20s and now could not attain. We had to choose between kids, houses, and time. Those with new cars had no houses; those with houses, no children; those with children, no houses. A few lucky supercouples—a lawyer married to a doctor, say, . . . had everything but time." Unable to fulfill the American Dream, the postmodern generation saved less, borrowed more, deferred parenthood, comforted itself with the luxuries that were marketed as symbols of style and distinctiveness, and generally surrendered to the hedonism of lives infused with extravagant details: imported mineral water, Godiva chocolates, sun-dried tomatoes, coarse-gain mustard, single-estate tea, shade-grown coffee, sea bass, and fresh-cut flowers on the table at all times; clothes by Issey Miyake, Perry Ellis, Hugo Boss, and Liz Claiborne; Krups coffee makers, Braun juicers, Kitchen Aid appliances, Tag Heuer watches, Leica cameras, $75 haircuts, $300 shoes; the Netherlands Antilles in January; a cottage by the sea in July; Beaujolais Nouveau in November. The signs—literally—of this new materialism appeared everywhere. Bumper stickers advising "Shop Till You Drop" or boasting "Who Dies with The Most Toys Wins" and "I'm Spending My Kids' Inheritance"; sweatshirts announcing "Dear Santa: I Want It All."

The culture of materialism that developed in the 1980s also brought new patterns of social behavior. Barbara Ehrenreich observed that the cornerstone of the "yuppie strategy" was the determination of both men and women to find proven salary earners as potential marriage partners. The result was the consolidation of an androgynous caste- and guild-like class fraction, characterized by the high educational status of both men and women and by very high household incomes. The importance of marrying suitably qualified partners, in turn, intensified the potency of material signifiers: "Since bank accounts and resumes are not visible attributes, a myriad of other cues were required to sort the good prospects from the losers. Upscale spending patterns created the cultural space in which the financially well matched could find each other—far from the burger eaters and Bud drinkers and those unfortunate enough to wear unnatural fibers."[10] More generally, with more married women established in the workforce, more households had the capacity for discretionary spending. And this meant that neighbors were no longer the principal point of reference. In the workplace, most employees are exposed to the spending habits of people across a wider economic spectrum, particularly those employees who work in white-collar settings. "As the workplace replaced the coffee klatch and the backyard barbecue as locations of social contact, workplace conversation became a source for information on who went where for vacation, who was having a deck put on the house, and whether the kids were going to dance class, summer camp, or karate lessons."[11]

The New Economy and Affluenza

The emerging "new economy," based on digital technologies, economic, and cultural globalization, and the growth of employment in finance, insurance, real estate, biomedical, and dot-com industries, was producing a "winner-take-all" society, with a big gap suddenly opening up between middle-middle and upper-middle income households.[12] A distinctive feature of the new economy was that higher-income earners had emerged in occupations that had only a weakly established social status. A "new bourgeoisie" quickly emerged, consisting of "symbolic analysts": economists, financial analysts, management consultants, personnel experts, designers, marketing experts, purchasers, and so on. They were joined by a "new petit bourgeoisie" dominated by well-paid junior commercial executives, engineers, skilled high-technicians, medical and social service personnel, and people directly involved in cultural production: authors, editors, radio and TV producers and presenters, magazine journalists, and the like.[13] Scott Lash and others refer to these classes as the "advanced services middle classes"—the innovative class fraction associated with "reflexive modernization" and the principal patrons of symbolic consumption.[14] Soja refers to them simply as "Upper Professionals," noting that this group "demands much more and has the public and private power to make its demands fit into the crowded, edgy, and fragmented built environment, increasingly shaping the city building process to their own image."[15]

Together, the new bourgeoisie and new petit bourgeoisie came to dominate the top of the hourglass economy, with salaries commensurate with their increasing economic influence. In 1975, the most affluent 20 percent of American households accounted for just over 43 percent of aggregate household incomes, while the top 5 percent accounted for just over 16 percent. By 2005, the numbers were 50.4 percent and 22.2, percent respectively. The average income of the top 20 percent of households in 2005 was $159,583, up from the 1995 average of $96,189 in constant (2005) dollars. The average income of the top 5 percent meanwhile jumped from $145,968 to $281,155.[16] These are the households that set the ever-rising standard for the Paradise Spell, living out the American Dream Extreme through relentless lifestyles of competitive consumption and grab-it-yourself materialism. Instructed by glossy lifestyle magazines like *Architectural*

Digest, Best Life, Cosmopolitan, Country Living, GQ, House Beautiful, Martha Stewart's Living, Pool & Spa Living, Stuff, Success, and *Trump Magazine,* America's upper-middle classes have developed chronic cases of "luxury fever" and "affluenza."[17] Incited by television programs like *Lifestyles of the Rich and Famous, MTV Cribs,* and *How I'm Livin'* (often celebrating the excesses of public vulgarians like Donald Trump, Conrad Black, and Paris Hilton), they learned to recalibrate the Paradise Spell ever-upward against the new metrics of attention-getting extravagance established by what Robert Frank calls "Richistan": the America occupied by the new super-rich.[18]

Advertising strategies, playing to the sensibilities and dispositions of the Paradise Spell, quickly shifted away from the simple iconology of mid-twentieth century campaigns (presenting products as embodiments of effectiveness and quality) to exploit narcissism (where products are portrayed as instruments of self-awareness and self-actualization), totemism (where products are portrayed as emblems of group status and stylishness), and covetousness (where products are flatly presented as emblems of exclusivity and sheer wealth). Carefully targeted to specific market segments via psychographics,[19] branding and niche marketing have become key strategies for advertisers, not least in reenchanting upscale suburban subdivisions as the preferred ecology for affluent households.

Consumers and Their Priorities

Whereas the democratic utopias of postwar suburbs were marketed principally to middle-middle and lower-middle-class households, the reenchanted suburbs of the contemporary New Metropolis are being marketed principally to the prosperous class fractions generated by the reshaped social and occupational structures of the "new economy." Broadly speaking, this group is coincident with the top quintile in terms of household income, but market researchers and business analysts know that households differ in their preferences and propensities. One of the most sophisticated approaches to consumer behavior is that developed by SRI Consulting Business Intelligence (SRIC-BI), which has developed a taxonomy of consumers based on market research into their values and priorities. According to SRIC-BI's VALS™ system of market analysis, consumers can be differentiated according to three kinds of primary motivations: ideals, achievement, and self-expression. Consumers who are primarily motivated by ideals are guided in their purchasing decisions by information and principles. Consumers who are primarily motivated by achievement look for products that demonstrate success to their peers. Consumers who are primarily motivated by self-expression have patterns of expenditure oriented toward social or physical activity, variety, and thrills.[20]

Consumers can also, of course, be differentiated in terms of their resources. Income and wealth are clearly important, but SRIC-BI's VALS™ taxonomy also takes into account people's resources in terms of health, self-confidence, energy, and awareness of current ideas, products, and styles. Taking both primary motivations and resources into account, the VALS™ typology includes eight market "segments" (Figure 7.1). Consumers whose primary motivations are ideals-based are subdivided into two segments: "Believers," who have fewer resources, and "Thinkers," who have more. Similarly, consumers whose primary motivation is self-expression consist of "Makers" (fewer resources) and "Experiencers" (more resources). Consumers whose primary motivation is achievement consist of "Strivers" (fewer resources) and "Achievers" (more resources). Two other distinctive segments are defined principally in terms of their resources: "Innovators," who have abundant resources and are able to indulge all three primary motivations

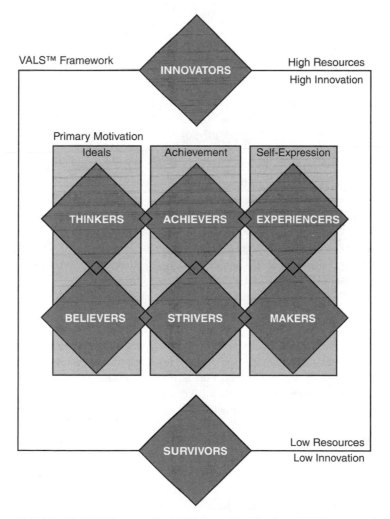

FIGURE 7-1. The VALS™ framework. ©SRI Consulting Business Intelligence. All rights reserved. (SRI-BC *http://www.sric-bi.com/VALS/types.shtml*)

to varying degrees; and "Survivors," with so few resources that they must focus on meeting needs rather than fulfilling desires, and are unable to express a strong primary motivation through their patterns of consumption.

In terms of the material culture of contemporary suburbia, the two most interesting segments are Innovators and Achievers.[21] Innovators tend to be well-educated, self-confident, open to innovation, and energetic. They tend to experience more "positive life experiences" (promotions, raises, etc.) than other groups and, when they do, they often reward themselves with some form of consumption. Their consumption patterns reflect cultivated tastes for upscale niche products and services. Achievers also have relatively high levels of resources but are characterized primarily by their conservatism and the emphasis that they place on status, structure, stability, and predictability. As consumers, they favor homes, neighborhoods, products, and services that demonstrate their success to their peers. Achievers are highly imitative, making purchases similar to those of others whose opinions they value or of those they wish to emulate.[22]

Together, Innovators and Achievers make up almost 25 percent of the adult population in the United States. Innovators account for one in ten of the adult population. They are, however, distributed unevenly across metropolitan areas. The pattern reflects, very broadly, the economic and cultural geography of the country (Figure 7.2). In the San Francisco–Oakland Metropolitan Statistical Area and the Washington, D.C.–Baltimore Metropolitan Statistical Area, around 20 percent of the adult population are Innovators. Other Metropolitan Statistical Areas where Innovators represent at least 15 percent of the adult population include Austin–San Marcos, Boston–Worcester–Lawrence, Charlottesville (Virginia), Corvallis (Oregon), Denver–Boulder–Greeley, Iowa City, Madison (Wisconsin), New London–Norwich (Connecticut), New York–Northern New Jersey–Long Island, Raleigh–Durham–Chapel Hill, Santa Fe (New Mexico), and Rochester (Minnesota).

All of these Metropolitan Statistical Areas except New York–Northern New Jersey–Long Island also contain above-average percentages of Achievers—that is, more than 14 percent of the adult population. In some Metropolitan Statistical Areas, Achievers constitute between 18 and 21 percent of the adult population: Anchorage, Atlanta, Austin–San Marcos, Cedar Rapids (Iowa), Colorado Springs, Denver–Boulder–Greeley, Fort Collins–Loveland (Colorado), Madison (Wisconsin), Minneapolis–St. Paul, Portland–Salem, Provo–Orem (Utah), Rochester (Minnesota), Salt Lake City, San Francisco–Oakland, and Seattle–Tacoma–Bremerton.

This broad geography also is apparent at the more detailed spatial scale of zip code areas. Concentrations of Innovators reflect a bicoastal pattern, the highest percentages occurring within metropolitan areas of the northeastern seaboard and coastal California. The highest concentrations of Achievers, in contrast, are distributed throughout America's heartland, almost all of them located in the suburbs of metropolitan areas. At the scale of five-digit zip codes, we can also begin to see the degree of residential segregation of different consumer groups within

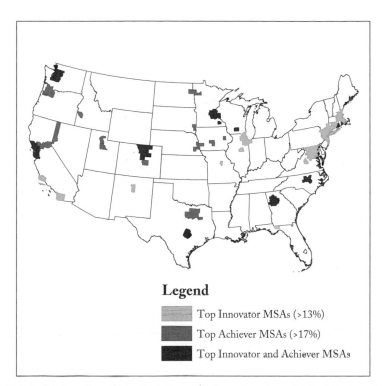

Legend

◻ Top Innovator MSAs (>13%)

▨ Top Achiever MSAs (>17%)

◼ Top Innovator and Achiever MSAs

FIGURE 7-2. Metropolitan Statistical Areas in the coterminous United States with the highest percentages of Innovators (13 percent or more of the adult population) and Achievers (17 percent or more of the adult population). (Paul L. Knox, "Vulgaria: The Re-Enchantment of Suburbia," *Opolis* 1 [2005], fig. 1, p. 39; data from SRI Consulting Business Intelligence's GeoVALS™)

metropolitan areas. In the Washington, D.C., metropolitan area, for example, high concentrations of Innovators reflect a distinctive social geography, dominating Fairfax County (Virginia), along with much of Montgomery County (Maryland) and Loudoun County (Virginia) (Figure 7.3). Achievers are localized in a broadly similar but rather more decentralized pattern (Figure 7.4). Concentrations of Innovators constitute more than 50 percent of the adult population in some zip code areas, reaching around 58 percent in Kenilworth (Cook County, Illinois), Glen Echo (Montgomery County, Maryland), and Waban (Middlesex County, Massachusetts). Achievers are somewhat less concentrated, with the highest levels at the scale of zip codes ranging between 30 and 35 percent.

By definition, Innovators are very sophisticated consumers of place. Along with Achievers, they can be expected to want houses that make a clear statement about themselves and their lifestyles (basically: "I've got a big/bigger/better equipped/more spectacular/more luxurious one"). But for resource-rich, successful, energetic, and aspirational consumers, a house is just the beginning. It must

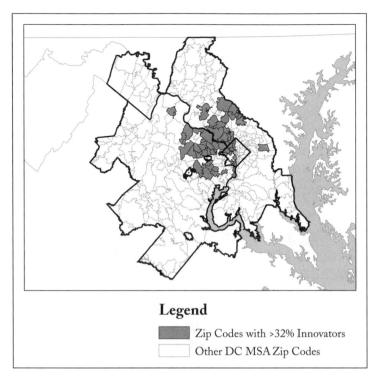

FIGURE 7-3. Distribution of zip code areas with a high incidence of Innovators (top quintile) in the Washington, D.C., Metropolitan Statistical Area. (Paul L. Knox, "Vulgaria: The Re-enchantment of Suburbia," *Opolis* 1 [2005], fig. 2, p. 41; data from SRI Consulting Business Intelligence's GeoVALS™)

also be a showcase for the right "stuff": the furnishings, possessions, and equipment necessary for the enactment of their preferred lifestyle and self-image.

The Right Stuff

Just what constitutes the right stuff is an increasingly important issue in a society where traditional institutions and class markers are being eclipsed by fragmented and segmented lifestyle groupings and where individuals, increasingly influenced by the Paradise Spell, have much greater expectations than ever before in expressing and creating their own individual identity. In this business of identity creation, the consumption of everything from houses to furniture and clothes is central, and the affluent new class fractions of the new economy have become a research and development lab for consumer preferences as well as the promoters of an intensified and voracious consumption ethic. Houses, neighborhoods, interior design, clothes, gadgets, food—everything—is now freighted with meaning. These meanings are shared among social sub-groups and become key markers of status, lifestyle, and identity. As Virginia Postrel puts it, "*I like this* becomes *I'm like this*."[23] Consumption is not simply an act of purchase in

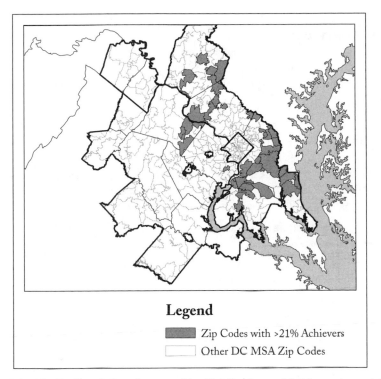

Legend

▨ Zip Codes with >21% Achievers

▢ Other DC MSA Zip Codes

FIGURE 7-4. Distribution of zip code areas with a high incidence of Achievers (top quintile) in the Washington, D.C., Metropolitan Statistical Area. (Paul L. Knox, "Vulgaria: The Re-enchantment of Suburbia," *Opolis* 1 [2005], fig. 3, p. 42; data from SRI Consulting Business Intelligence's GeoVALS™)

pursuit of needs and wants but a social process in which different groups relate to specific goods and artifacts in complex ways, deploying symbolic languages of exclusion, entitlement, and distinction: identity kits for fragmented class fractions and specialized lifestyle groupings. Material consumption has evolved from its mid-century role as the corollary to mass production to a role as a cultural act; and now to a role in the fulfillment of dreams, images, and pleasures.[24]

The era of mass society, "in which everyone likes the same, reads the same, practices the same, was a short intermezzo. . . . What has . . . emerged is not the standardization and unification of consumption but rather the enormous pluralization of tastes, practices, enjoyments and needs. The quantity of money available for spending continues to divide men and women but so do the kinds and types of enjoyment, practices, pleasure which they seek. . . . More importantly, the different patterns of consumption have become embedded in a variety of lifestyles."[25] This differentiation makes for a sociocultural environment in which the emphasis is not so much on ownership and consumption per se but on the possession of particular *combinations* of things and the *style* of consumption. In addition, the space-time compression of the new economy and the associated globalization and homogenization of popular culture has fostered the perceived need for distinctiveness and identity: "This society which eliminates geographical distance reproduces distance internally as spectacular separation."[26] In Debord's "society of the spectacle," where the emphasis is on appearances, the symbolic properties of urban settings and material possessions have come to assume unprecedented importance.

Houses and Consumption.   Space and place are crucial elements of consumer identities, and people's homes are central to both their self-perception and their social standing.[27] The definition of home and, by extension, consumption for the home, carries a heavy ideological burden. John Archer, in *Architecture and Suburbia*, lists the following ideological principles and premises embedded in single-family suburban homes, per the American Dream:

> *self*: the dwelling is understood as a principal instrument for self-realization.
>
> *property*: private property is a fundamental condition of selfhood, and of the encompassing political and economic systems.
>
> *identity*: the dwelling serves as an apparatus for the articulation of identity and belonging.
>
> *privacy*: the opportunity for privacy is central to the articulation of identity.
>
> *pastoral*: the pastoral ideal is part of the mythic energy of suburbia.
>
> *family unit*: the principal paradigm of American housing remains the detached single-nuclear-family dwelling, which reciprocally defines the family, . . . [and]
>
> *home-as-castle*: despite, or perhaps because of, the increased complexity of relations between individuals and corporate, institutional, national, and global entities, the dwelling remains almost exclusively centered on the individual and the family, and its perimeter increasingly serves as a defense against all other forces.[28]

To these we can add the convention of the dwelling as a display cabinet for material goods and as a stage for the enactment of privatized lifestyles: backyard barbecues and relaxing-by-the-pool; front yard manicuring and car-washing; and web surfing, home theater viewing, and occasional dinner parties indoors. In master-planned developments we can add the prescribed aesthetic and behavioral conventions imposed through servitude regimes.

Social distinctions, previously marked by the ownership of particular items of consumer durables, now have to be established via the symbolism of aestheticized settings and commodities. "Surrounded by our things, we are constantly instructed in who we are and what we aspire to."[29] Patterns of home-related consumption are among the most powerful and pervasive social processes within metroburbia. They are epigrammatic, able to carry symbolic meanings; they mold people's consciousness of place, help people to construct real places and to connect the key realms of nature, social relations, and meaning.[30] Much of this power is a function of the web of signification ascribed to particular commodities and groups of commodities and the settings—homes and neighborhoods—in which they are stored and displayed. "People want their stuff to tell them who they are. They ask that inanimate objects serve as stand-ins for deeper qualities. Not just pretty flowers but a built-in serenity is taken to exist in a Laura Ashley wallpaper pattern. Not just style but the character of a person is presumed to be made manifest by a Ralph Lauren blazer. People disappear into their clothes."[31] Signs and symbols "reflect and refract reality. Social life is impregnated with signs that make it classifiable, intelligible, and meaningful."[32] Each signifier, whether it is a house or a sofa, an automobile or kitchen cabinetry, can be ascribed not only a denotative, surface-level meaning but also one or more second-level, connotative meanings.

Even modestly affluent households are sophisticated and reflexive, highly adept at the art of positional consumption. But people's values and priorities as consumers tend to vary significantly across relatively narrowly differentiated bands of income and wealth, reflecting different lifestyle preferences. Consumers select houses and purchase products, services, and experiences that give shape, substance, and character to their particular identities and lifestyles. Some, for example, will be more predisposed to bigness, bling, and the themed and gated developments at the leading edges of the New Metropolis than will others. Michel Maffesoli argues that, in many instances, this sort of predisposition results in affectual or "neotribal" groupings of people who feel and think alike as well as sharing the same tastes and preferences for material goods.[33]

For every group, the symbolism and meaning of material goods and the built environment is under constant construction and reconstruction, interpretation and reinterpretation. New products, new designs, and shifts in taste and style have the tendency to exclude those who may not be "in the know" or do not have the means to make "necessary" changes to their ensemble of possessions and patterns of activities. At the level of the individual household, it is important to maintain a consistent aesthetic as new objects are incorporated and the old discarded. In marketing terms this is known as brand coherence. Its significance was recognized long ago by the French philosopher Denis Diderot, and it is sometimes referred to as the "Diderot effect." Diderot had been working quite happily in his crowded, chaotic, and rather shabby study until he received a fancy velvet smoking jacket as a gift. He liked his new jacket but soon noticed that its quality made his surroundings seem threadbare. His desk, rug, and chairs looked scruffy by comparison. So, one by one, he found himself replacing his furnishings with new ones that matched the jacket's elegant tone. He realized (though he later regretted it) that he had felt the need for a sense of coherence, a sense that nothing was out of place. Modern marketing and advertising is geared to exploit the Diderot effect, seeking to stimulate continuous rounds of purchasing that encourage people to maintain a cultural consistency in their constellations of consumer goods. In addition to stimulating wants, advertising assumes a key role in teaching people how to dress, how to furnish a home, and how to signify status through ensembles of possessions that amount to "brandscapes." Targeted via psychographics and geodemographics, market segments are not only identified but reinforced through selective and sophisticated advertising strategies that are pitched to their particular sensibilities and dispositions. It all becomes something of a self-fulfilling prophesy.

Habitus.   Members of particular market segments, neotribal groups and class fractions develop collective perceptual and evaluative schemata—cognitive structures and dispositions—that derive from their everyday experience. These schemata operate at a subconscious level, through commonplace daily practices, dress codes, use of language, and comportment, as well as patterns of consumption. The result is what French social theorist Pierre Bourdieu has conceptualized as "habitus": a distinctive set of structured beliefs and dispositions "in which each dimension of lifestyle symbolizes with others."[34] Habitus incorporates both habit and habitat. It frames the sense of one's place in both the physical and social senses: it "implies a 'sense of one's place' but also a 'sense of the other's place.' For example, we say of an item of clothing, a piece of furniture or a book: 'that's petit-bourgeois' or 'that's intellectual.' "[35] Or "That's me," "That's you," or "That's vulgar."

Habitus exists in different "fields" of life. Bordieu's concept of field refers to subject-areas (religious, legal, political, scientific, artistic, academic, sociological, market-oriented, for example) with their own logic and with their own valuing of objects and practices. The subject area of housing may be viewed as a market-oriented field characterized by questions of what constitutes a desirable house, how different rooms are used, how it should be furnished, how big it should be, and so on. People's homes are an important means of reaffirming and delineating class cultures—they are what Bourdieu referred to as "structuring structures." In this sense, housing (especially single-family detached suburban dwellings) represents a durable framework that serves to sustain lifestyles in accord with owners' dispositions. Other market-oriented fields within which class fractions develop a distinctive habitus include food, sport, and clothes. Habitus and field, then, are mutually interdependent, as a habitus always exists in relation to a given field. Fields are arenas of conflict and competition; habitus provides the players in each field with a sense of how to keep score.

Each class fraction (or market segment or neotribal group) establishes, sustains, and extends a habitus by exercising its economic, social, cultural, and symbolic capital. Economic capital is any form of wealth that is easily turned into money. Social capital derives from the social connections of family, place, and neighborhood, often inherited through class membership. Cultural capital is the sum of credentials, skill, and knowledge acquired through education and upbringing. Symbolic capital—really a sub-set of cultural capital—derives from the command of superior taste. In traditional Western social hierarchies, superior taste is the distinguishing attribute of the habitus of bourgeois class fractions, whose members are socialized to appreciate fine art and music, to like certain foods, to understand complicated art forms, and to master certain context-specific manners, vocabularies, and demeanors.

Different class fractions and market segments have different combinations of economic, social, cultural, and symbolic capital. Traditional upper-class groups command high levels of each kind of capital; working-class groups have little of any kind; the *nouveau riche* have plenty of economic capital but less social, cultural, and symbolic capital; and so on. The point here is that these combinations of different kinds of capital result in different dispositions: habitus. Hence, for example, the more refined consumption practices of the traditional upper classes, in contrast to the conspicuous consumption of the *nouveau riche*, as noted by Thorstein Veblen in 1899.[36] In the more contemporary context of market segments, Survivors have little economic, social, cultural, or symbolic capital, while Innovators have ample amounts of each. Achievers have fairly high levels of economic capital, but can be expected to have rather less by way of social and cultural capital and less still by way of symbolic capital.

The fact that symbolic capital is vulnerable to shifts in the denotative and connotative meaning of goods and practices and to the availability of new product lines and new styles only makes it more potent as a measure of distinction. The habitus of dominant class fractions is inevitably undermined by the accessibility and popularization of goods and practices that were formerly exclusive, so that the process of maintaining habitus is continuous. The signs and symbols of distinction have constantly to be shuffled, inverted, or displaced. New languages of taste and identity have to be mastered. New products have to be inducted as desirable or enchanting (or not); the once-fashionable has to be condemned as dated or tasteless; kitsch has to be consecrated as cool; and showiness has to be cultivated as a desirable trait. In this context, the cultural capital available to a class fraction is critical. Cultural capital is not simply equivalent to cultural literacy but is, rather, a "feel for the game," a product of knowledgeability of the symbolic meaning of particular cultural artifacts and sociocultural practices. Such competence comes easily to members of the new bourgeoisie and new petit bourgeoisie who are designers, marketing experts, or those directly involved in cultural production—individuals who are likely to be Innovators, in terms of market segments. Achievers, Thinkers, and Experiencers would be expected to have rather less cultural capital, and for it to be specialized according to their primary motivation. Experiencers, for example, will tend to have a relatively high level of cultural capital in relation to sports, outdoor recreation, fashion, entertainment, social activities, and having "cool" stuff. Achievers' cultural capital will center around knowledgeability of established, prestige products, services, and behaviors that demonstrate success to their peers: things that may be seen by others as crass or vulgar.

Landscapes and Brandscapes: Consumption and Social Ecology

The social ecology of the mid- and late twentieth-century city could be mapped and analyzed through a mixture of census-based economic, demographic, and housing statistics. The striking finding by geographers and sociologists was that the sociospatial organization of most cities resulted in wedge-shaped sectors of socioeconomic status, concentric zones of neighborhoods with distinctive demographic attributes (age, household structure, etc), and isolated clusters of neighborhoods with distinctive ethnicities.[37] In contrast, the social ecology of the New Metropolis is a fragmented mosaic, differentiated into neighborhoods that reflect the segmented lifestyle groupings and class fractions of postindustrial, postmodern society, each with different degrees of access to economic, social, cultural, and symbolic capital.

Mapping and analyzing this mosaic requires detailed information about the patterns of consumption that are so important to status and identity in contemporary America. Consumer research consultants, using multivariate statistical methods similar to those used by urban social geographers to tease out the spatial structure of twentieth-century cities, have identified the matrix of contemporary household types in America according to their distinctive consumption patterns as well as their socioeconomic and demographic attributes and their typical residential settings. The best-known and most comprehensive classification of landscapes and brandscapes is the Nielsen corporation's Claritas PRIZM®NE system.[38] It is based on block-level data from the 2000 U.S. census of population, merged with consumption and lifestyle data from list-based records for nearly 200 million households, along with detailed profiles of more than 890,000 households from sources such as R. L. Polk's new car buyers and Simmons lifestyle surveys. The PRIZM system crunches these data into fourteen major sociogeographic groups: Urban Uptown, Midtown Mix, Urban Cores, Elite Suburbs, Affluentials, Middleburbs, Inner Suburbs, Second City Society, City Centers, Micro-City Blues, Landed Gentry, Country Comfort, Middle America, and Rustic Living. Each of these, in turn, is subdivided, giving a total of sixty-six distinctive market segments, each numbered according to socioeconomic rank (taking into account characteristics such as income, education, occupation, and home value). At least twenty-five of these segments relate to the suburbs, exurbs, and satellite centers of the New Metropolis.

In this classification, the most affluent suburban social group, Elite Suburbs, accounted for just over 5 percent of all U.S. households in 2006: roughly 5.8 million households. Households in the most exclusive market segment within this group (and ranked #1 in the whole country) are Upper Crust households, dominated by empty-nesting couples over fifty-five years old. No segment has a higher concentration of residents earning over $200,000 a year or possessing a postgraduate degree. Typical aspects of this segment's consumption profile include expenditures of $3,000 or more on foreign travel each year, shopping at Bloomingdale's, subscriptions to *Atlantic Monthly* magazine, watching the Golf Channel, and driving Jaguar XKs. Blue Blood Estates (ranked #2 overall) account for a further one million households (about 1 percent of the total) and are characterized by married couples with children, college degrees, a significant percentage of Asian Americans, and six-figure incomes earned by business executives, managers, and professionals living in million-dollar homes with manicured lawns, driving high-end cars like the Audi A8, reading *Architectural Digest* and *Scientific American*, and joining exclusive private clubs. Movers & Shakers (#3; 1.8 million households; 1.6 percent of the total in 2006) represent America's up-and-coming business class who live in a wealthy suburban world of dual-income couples who are highly educated, typically between the ages of thirty-five and fifty-four, often with children. Typical aspects of their segment's consumption profile include scuba diving or snorkeling, eating at Bertucci's, reading *Inc.* magazine, and driving Porsche 911s. The Winner's Circle segment (#6; 1.2 million households; 1.1 percent) is characterized by affluent younger households: mostly twenty-five- to thirty-four-year-old couples with large families in new-money subdivisions. Surrounding their homes are the signs of upscale living: recreational parks, golf courses, and upscale malls. With a median income of nearly $190,000, Winner's Circle residents are big spenders who like to travel, ski, go out to eat, shop at clothing boutiques like Ann Taylor, and drive Infiniti SUVs.

The next most affluent group is dubbed Affluentials by Claritas. They number more than 6.5 million households (7.7 percent) and fall into six segments with distinctive suburban lifestyles—Executive Suites (#8), New Empty Nests (#14), Pools & Patios (#15), Beltway Boomers (#17), Kids & Cul-de-Sacs (#18), and Home Sweet Home (#19). The median home value for the group as a whole in 2006 was about $200,000, and most couples in this social group tend to have college degrees and white-collar jobs. Affluent exurbs and boomburbs are the settings for the five market segments of the Landed Gentry group (10.2 million households; 9 percent)—Country Squires (#5), Big Fish, Small Pond (#9), God's Country (#11), Fast-Track Families (#20), and Country Casuals (#25). Many of the households are baby-boomer families with professional jobs and expansive homes. They are twice as likely as average Americans to telecommute. With their upscale incomes, they can afford to spend heavily on consumer electronics, wireless and computer technology, luxury cars (Lexus SUV, Lexus LS430, Toyota Land Cruiser) powerboats, country club membership and sports (golf, tennis, swimming), books and magazines (*Atlantic Monthly, Family Fun, Skiing*), children's toys, and exercise equipment. Another exurban segment is Greenbelt Sports (#23; 1.6 million households; 1.4 percent; part of the Country Comfort group), distinctive for its residents' active lifestyles featuring skiing, canoeing, backpacking, boating, and mountain biking. Most of these middle-aged residents are married and college-educated and own new homes; about a third have children.

Figure 7.5 shows the location of the dominant PRIZM®NE sub-groups in the Louisville, Kentucky, metro area. Consistent with the fragmented and mosaic form that is typical of contemporary American metropolitan regions, affluent suburban sub-groups are the principal market segment in zip code areas at every point of the compass in the Louisville metro region as well as close to the city center. The eastern and northeastern part of the region, between 7 and 20 miles from downtown, is blanketed by affluent suburban lifestyle groups,

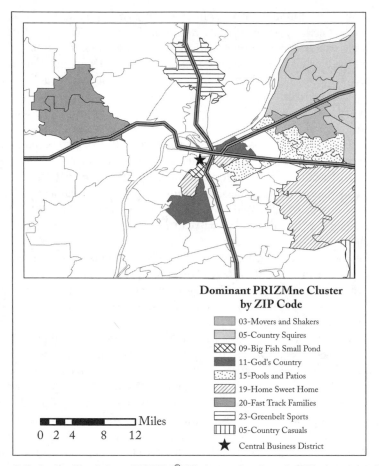

FIGURE 7-5. Louisville: Selected PRIZM®NE clusters by zip code. (Claritas, a service of the Nielsen Company)

including the Louisville region's only representative of the Elite Suburbs social group, Movers and Shakers. The Affluentials social group is represented in these eastern suburbs by Pools and Patios and Home Sweet Home market segments, while the suburbs and exurbs to the farthest northeast and southeast of the metro region are dominated, respectively, by Country Squires and Country Casuals households, both part of the Landed Gentry social group. Closer to the downtown are older-established neighborhoods dominated by Home Sweet Home lifestyle sub-group, the God's Country sub-group, and the Big Fish Small Pond sub-group. To the west of the central city, between twelve and fifteen miles from the downtown, are exurbs dominated by Fast Track Families, part of the Landed Gentry social group. Finally, the exurbs to the north of the region, between seven and fifteen miles from downtown Louisville, are dominated by Greenbelt Sports households, part of the Country Comfort social group.

Lifestyle Enclaves and Community

The residential mosaic revealed by market segment analysis reflects the secession of the successful into lifestyle enclaves that are accommodated in the master-planned developments and upscale subdivisions of the New Metropolis. It is, in simple terms, a reflection of demand and supply, mediated by focus groups and advertising. On the demand side, there is an implicit expectation that secession brings with it a sense of community or communality through voluntary segregation of like-minded households. On the supply side it is explicit: a central feature of advertisements for master-planned and neotraditional developments and a foundational assertion of new urbanist design. "Community" is a key sales feature in the consumer housing industry, promised as part of every package of amenities by developers. "Community" is also central to new urbanist lore: on the one hand as the focus of the critique of conventional subdivisions (they are antithetical to community), and on the other as the focus of new urbanist market appeal (new urbanism fosters a sense of community).

Nevertheless, there is more to the secession of the successful than the quest for lifestyle communities or the (supposed) warm communality of traditional small-town settings. Richard Sennett has emphasized the primacy, in contemporary society, of avoidance of exposure to "otherness," noting how cities are organized either around spaces limited to (and carefully orchestrating) consumption; or spaces limited to (and carefully orchestrating) experience. " 'Exposure' more connotes the likelihood of being hurt than of being stimulated. . . . What is characteristic of our city-building is to wall off the differences between people, assuming that these differences are more likely to be mutually threatening than mutually stimulating. What we make in the urban realm are therefore bland, neutralizing spaces, spaces that remove the threat of social contact."[39] What Sennett is talking about here has more to do with exclusionary social segregation than the propagation of lifestyle and the quest for community. Participating in this voluntary segregation, he argues, people draw themselves a picture of who they are, an image that elides anything that might convey a feeling of difference or dissonance: a "purification ritual" that produces a myth of community solidarity.

As sociologists and social geographers have long been aware, the spatial segregation of sociodemographic or sociocultural groups does not necessarily equate to "community." Communities can only be said to exist where a degree of social coherence develops on the basis of mutual interdependence, which in turn produces a uniformity of custom, taste, and modes of thought and speech. Communities are "taken-for-granted" worlds of values, dispositions, and behaviors shared by social groups that may or may not be locality-based. Communality is something more: a form of social association and identity based on affective bonds. It is community experience at the level of consciousness. It requires an intense mutual involvement that is difficult to sustain in an affluent, faster-paced society and so only appears under conditions of stress: "crisis communality" that appears in suburban neighborhoods at times when there is an unusually strong threat to territorial exclusivity, amenities, or property values. NIMBY responses often generate communality, but only for the duration of the perceived threat.

According to classic sociological theory, communities should not exist at all in cities; or, at best, only in a weakened form. This idea first entered sociological theory in the nineteenth century by way of the writings of Ferdinand Tönnies, who argued that two basic forms of human association could be recognized in all cultural systems.[40] The first of these, *Gemeinschaft*, he related to an earlier period in which the basic unit of organization was the family or kin-group, with social relationships characterized by depth, continuity, cohesion, and fulfillment. The second, *Gesellschaft*, was seen as the product of urbanization and industrialization that resulted in social and economic relationships based on rationality, efficiency, and contractual obligations among individuals whose roles had become specialized. This perspective was subsequently reinforced by the writings of sociologists such as Durkheim, Simmel, Sumner, and Wirth, and has become part of the conventional wisdom about city life: it is not conducive to "community," however it might be defined. This view has been characterized as the "community lost" argument.[41]

Yet writers have long portrayed the city as an inherently human place, where sociability and friendliness are a natural consequence of social organization at the neighborhood level. Herbert Gans, following his classic 1962 study of the West End of Boston, suggested that we need not mourn the complete passing of the cohesive social networks and sense of community associated with village life because, he argued, these properties existed within the inner city in a series of "urban villages."[42] This perspective has become known as "community saved." Urban villages are most likely to develop in long-established working-class areas with a relatively stable population and a narrow range of occupations. The importance of permanence and immobility in fostering the development of local social systems has been stressed by many writers. The relative immobility of the working classes (in every sense: personal mobility, occupational mobility, and residential mobility) is a particularly important factor. Immobility results in a strengthening of vertical bonds of kinship as well as the horizontal bonds of friendship.

In contrast to the close-knit social networks of the urban village, the suburbs, populated by relos and upwardly mobile households, are seen by many observers as the antithesis of community. Lewis Mumford, for example, wrote in 1940 that the suburbs represent "a collective attempt to lead a private life,"[43] a view that was endorsed by a number of early sociological studies of suburbs.[44] Nevertheless, as Herbert Gans showed in his classic study of Levittown, some suburban neighborhoods do contain localized social networks with a degree of cohesion.[45] Suburban neighborhoods can be thought of as conditional communities: "communities of limited liability"—one of a series of social networks in which individuals may choose to participate. This view has been translated as "community transformed." Instead of urban communities breaking up, they can be thought of as breaking down into an ever-increasing number of independent networks, only some of which are locality-based. These consist mainly of social networks related to voluntary associations of various kinds: parent-teacher associations, gardening clubs, country clubs, rotary clubs, and so on. Such networks are not really communities—they lack the solidity and lasting social bonds that result in cohesiveness and a uniformity of custom, taste, and modes of thought and speech—though they do provide some foundation for the potential formation of genuine community.

Community Lost (Again)

In the New Metropolis, it seems, secession is its own punishment. Americans are lonely, and getting lonelier. Even close personal networks are shrinking. In 1985, the General Social Survey collected the first nationally representative data on the confidants with whom Americans discuss important matters; the average number of confidants per person was just three. But by 2004 it had fallen to two, and the number of people reporting that they had no one at all with whom they discuss important personal matters nearly tripled. Those who did have confidants typically had to rely more on family members and less on friends and contacts made through voluntary associations and neighborhoods. The number of people who said they could count a neighbor as a confidant dropped by more than half, from about 19 percent to about 8 percent.[46]

Although these surveys were drawn from the entire U.S. population, it seems reasonable to infer that the results apply to the suburban and exurban residents of metroburbia. In fact, they probably understate the situation. It has been the boomburbs, new suburbs, and exurbs of the New Metropolis, after all, that have been the leading edges not just of demographic change in America but also the leading edges of sociocultural change: the "militant parochialism" of the New Right, along with its economic fundamentalism and moralizing social conservatism.[47] In contrast to the advertising hype, the private master-planned developments that dominate the leading edges of the New Metropolis do not in fact seem conducive to a sense of community, especially in the incarcerated neighborhoods that are gated and walled. Indeed, the idea of a gated "community" is something of an oxymoron. A national survey of residents of homeowner associations found that, while most respondents reported that their developments were "friendly," only eight percent felt they were "neighborly and tight-knit," while 28 percent felt that their neighborhood was "distant or private" in feeling.[48]

Robert Putnam captured the isolation of contemporary suburbia in his influential book, *Bowling Alone*. Putnam argues that Americans' social connectedness and civic engagement reached a peak between 1945 and 1965 and then declined precipitously. On average, Putnam found, involvement in political and community events declined by about 25 percent between 1973 and 2000. This includes declines in presidential voting as well as declines in local civic engagement such as serving on town committees, local organizations, and clubs. Levels of basic socializing also dropped, as did the general level of mutual trust. The number of Americans attending club meetings declined by 58 percent, family dinners declined 33 percent, and having friends over to one's home declined by 45 percent.

Putnam ascribes these changes to a combination of factors, including generational change, the pressures of maintaining dual-career households, and the growth of electronic entertainment: 500-channel cable and satellite television, VCRs, DVDs, electronic games, the Internet. Another significant factor, he concludes, is suburbanization. The spread of people's daily lives across work, shopping, and errands in the decentralized and splintered New Metropolis tends to pull apart any community boundedness that may exist. Ever-longer commutes mean that the amount of time spent driving alone has reduced not only the amount of time available for socializing and civic engagement but also people's energy and inclination to do so. Putnam also suggests that the fragmentation of the leading edges of suburbia into "enclaves segregated by race, class, education, life stage, and so on" undermines rather than sustains community. The homogeneity of these enclaves, he argues, reduces the local disputes and issues that draw neighbors into public contact.[49]

Globalization has not helped, either. The global diffusion of material culture, pop culture, celebrity culture, sports culture, and travel culture has drawn more of people's time and attention away from local issues and broadened their points of reference as consumers still further. Yet there is a dialectic here. As noted before, the more that people are engaged in globalized culture and generalized lifestyles, the more that local identities are valued and the more people feel the need for familiar and centered settings with a sense of identity. This, of course, is an important basis of the rationale advanced by advocates of traditional neighborhood development and new urbanism. It is also a need recognized by builders and developers, whose sales pitches and advertising copy routinely highlight "community" as a feature of their products.

Community Commodified

Because it is so much in demand, and yet so elusive, a sense of community is an attribute that is claimed by marketers of all sorts of commodities. Internet marketers apply the label "community" to random collections of website visitors. In popular usage, we speak and read quite routinely of communities of business people, of people of a particular age group, or of specialized interests. In marketing copy, "community" is a ubiquitous and versatile touchstone (along with speed, size, novelty, and convenience), relentlessly referenced in relation to all kinds of products and services, from cell phones to banking. In the process, of course, the term "community" itself is devalued and debased; no longer a sociological concept, only an advertising cliché.

For many builders and developers, "community" has come to be equated simply with residential development: a tract of housing, a subdivision, a resort, or a condominium project. For design professionals, and new urbanists in particular, propagating a sense of community is a central mission. Emily Talen, a planning academic who is generally positive about new urbanism, notes that new urbanism "lives by an unswerving belief in the ability of the built environment to create a 'sense of community.' " Better, she observes, to make claims in terms of sense of place or the common good. That way, new urbanism is less vulnerable to charges of spatial determinism.[50] The same is true for other approaches to urban design, almost all of which make claim to fostering a sense of community. Empirical studies of people's perceived sense of community reveal mixed results, at best, in support of the claims for design in fostering such feelings.[51] Jill Grant, who has undertaken the most comprehensive and even-handed review of new urbanism, concludes that "while new urbanism certainly creates attractive places where people can meet, its bolder claims of generating community are not supported. If new urbanist projects do engender commonality amongst their residents, they do so primarily because of the homogeneity of those who buy into the projects. Residents self-select for their commitment to the principles that new urban projects reflect."[52]

But, like builders and developers, designers and planning consultants are selling a product, and "community," in spite of being hollowed out in terms of its sociological meaning, still has a lot of commercial resonance. As a result, bold claims about creating a sense of community are not renounced or qualified but, rather, hyped. As Jill Grant notes in relation to practitioners of new urbanism, "Their success in becoming lucrative design practices reflects their salesmanship as well as their considerable expertise and innovation. At times they have created 'histories' for the places they propose, to add flesh to the skeletons of the urban form. They have taken clients and planners on tours of the kinds of places they seek to emulate, to give a visceral feel for the anticipated benefits. For the most part, though, they use watercolours, paintings, 3-D visualizations, and drawings as sentimental marketing tools to persuade potential clients of the power of their concepts."[53] The Congress for The New Urbanism has been particularly aggressive in marketing claims, organizing intense media coverage, offering workshops, videos, tool kits, brochures, and courses. The new urbanism division of the American Planning Association has adopted a similar proselytizing role, offering members advice on how to promote new urbanism and smart growth as vehicles for an improved sense of community and sense of place through the selective use of imagery, and how to spin potentially unpalatable aspects of the manifesto (e.g., reduced automobile use) into more marketable slogans ("increased choices in transportation"). In the end, it is hard not to conclude that the marketing and commodification of "community" is simply part of the broader sweep of consumerism and the increasing importance of the built environment as a stage and storehouse for material culture.

CHAPTER 8

Vulgaria

MORAL LANDSCAPES AT THE LEADING EDGES
OF THE NEW METROPOLIS

Previous chapters have shown how the suburban settings at the leading edges of the New Metropolis have been shaped by the interaction and interdependence of a broad set of actors that together constitute the "structures of building provision": intellectuals, designers, developers, builders, realtors, private governments, and households. These new suburbs, boomburbs, and exurbs, the most recent manifestation of a long sequence of enchantment, disenchantment, and reenchantment, have taken on a distinctive form, dominated by a combination of bigness, luxury, and neotraditional iconography. They are the physical settings in which the social relations and material culture of America's most affluent and influential class fraction are constituted, constrained, and mediated. Their significance, however, goes well beyond their role as settings for the daily lives of their residents. They provide landscapes that consist not only of the built environment but also an ensemble of material and social practices and symbolic representations with a "structured coherence"[1] that resonates across American society.

Our landscapes are uniquely revealing about who we are as a nation. Landscapes are mirrors of society: Our economy, politics, and culture are all inscribed in fields and fences, skyscrapers and suburbs. New England townscapes symbolize an idealized, family-centered, church-going, industrious, thrifty, and democratic community. Rust Belt townscapes are witnesses to our inability to manage local economies in the face of global economic competition. The landscapes of suburban commercial strips are microcosms of consumer culture; and so on. As palimpsests of economic and social history, landscapes not only echo and embody the fortunes of successive generations of inhabitants, they also reflect our individual behavior and even our ability to think and act collectively. Our daily surroundings are powerful but stealthy backdrops that can naturalize and reinforce dominant political and economic structures as if they were simply given and inevitable.[2] Laden with layers of symbolic meaning, everyday landscapes—including the people who inhabit them, their comportment, their clothes, their "stuff"—amount to moral geographies that both echo and tend to reproduce a society's core values and perform vital functions of social regulation.[3] As Kim Dovey observes:

> The more that structures and representations of power can be embedded in the framework of everyday life, the less questionable they become and the more effectively they can work. This is what lends built form a prime role as ideology. It is what Bourdieu calls the 'complicitous silence' of place as a framework to life that is the source of its deepest associations with power. . . . Ideology constructs place experience and design process at all levels as a necessary framework of belief—about the 'good life,' the 'nice house,' property, human rights, identity, privacy, family, and the individual. These are also beliefs about the state, authority, justice, democracy, class, status, gender, race, efficiency, and the public interest. The built environment is a primary medium for the techniques of establishing, legitimizing, and reproducing ideology at every scale from the house to the city. While 'ideology' has a traditional meaning linked to 'false consciousness,' it also has a broader meaning as a necessary relationship between consciousness and the structures of the material world. . . . As such, ideology is integrated with the 'web of meaning' we call culture.[4]

Don Mitchell notes that landscape is both an object and a social process, at once solidly material and ever changing: "Landscape is best understood as a certain kind of produced, lived, and represented space constructed out of struggles, compromises and temporarily settled relations of competing and cooperating social actors."[5] Landscape can also be thought of as a form of discourse. Places tell us stories, and we read them as spatial text. Most commentators, following Sharon Zukin, have interpreted this in terms of power relations. There is, of course, a complex nexus between built form and power, involving the entirety of the structures of building provision. Ultimately, though, placemaking is an inherently elite practice, determined by those in control of resources. Nigel Thrift has meanwhile noted that the literature in urban studies generally tends to foreground signification at the expense of affect—the feelings, emotions, and moods evoked in different settings as new developments change the spatial grammar of power. Cities, he points out, "are sources of numerous affective intensities which surge through them in ways we do not really understand but which form a constantly moving map that we can ill afford to ignore."[6] In the context of contemporary metroburbia, then, we should ask what sort of affect is generated and mobilized by McMansions, luxury subdivisions, private governments, walls, gates, security cameras, golf courses, country clubs, and megamalls.

Schlock-and-Awe Urbanism: Learning from Las Vegas

In a consumer economy that is driven by dreams and fantasies (Colin Campbell's "romantic capitalism"; David Brooks's "Paradise Spell")[7] and polarized by free-enterprise neoliberalism, the commodified reinterpretation of utopian ideals has created distinctive landscapes. These landscapes are not only emblematic of the materialism of U.S. consumer culture but also both a product and a cause of it. The landscapes of upscale suburban America are, more than anything else, ostentatious and meretricious. Infused with the bigness and bling endorsed by developers' focus groups, the private master-planned developments at the leading edges of the New Metropolis amount to what I have called schlock-and-awe urbanism.[8]

By any account, the capital of schlock-and-awe urbanism must be Las Vegas—the city of over-the-top spectacle, where the sights of Paris are just across the street from the canals of Venice, and right down the block from the Brooklyn Bridge. Las Vegas has been in the business of place marketing from the beginning. It is the apotheosis of Debord's "society of the spectacle" and Harvey's "degenerative utopias."[9] In the 1970s we were urged to "learn from Las Vegas" for the symbolism of its architectural forms, for the malleability of its visage, and for its significance as "The Great Proletarian Cultural Locomotive."[10] In fact, what other cities learned from Las Vegas was the power of the carnivalesque in economic redevelopment, the potency of gambling, and the importance of novelty and luxury.[11] Begun as "gateway to the Boulder Dam," briefly celebrated as "The Atomic City," and then famed as a "gangster diaspora" and "mobster metropolis" built with shoeboxes full of cash from the Mob and loans from the Teamsters' pension funds, Las Vegas has long understood the power of reinvention. During the 1950s, the town was the place to sin—a counterpoint to the family-oriented morality codes of the democratic utopias of post-war suburbia. By the 1970s, however, Las Vegas casinos' cabaret acts had come to seem passé in comparison to headliner acts that could sell out stadiums and arenas; while the city's saucy topless revues had come to seem both dated and sleazy. In 1976, Las Vegas lost its virtual monopoly over casino gambling when the state of New Jersey licensed Atlantic City as a gaming center.

Luckily for Las Vegas, developer Steve Wynn came on the scene. Wynn took a long-range perspective, studied market trends, and developed a strategy using junk-bond financing to build fantastic attractions and themed hotels with an unprecedented level of bigness and bling. First came the Mirage, which at $637 million, twenty-nine stories, and three million square feet, became the world's largest casino. Its principal features included an ecologically "authentic" tropical rain forest in a nine-story atrium; a 20,000-gallon marine fish tank behind the reception desk; and, outside, a fifty-four-foot volcano that spews steam and flames into the night sky for three minutes every half hour. The spectacular circus act of Siegfried and Roy (which featured white tigers, an almost extinct creature) topped off the spectacle by harnessing American pop culture allure with an extravagant show of what money could buy. In the first six weeks of operation, some 400 people had already bet one million dollars or more during a single visit, generating profits for the house that would spur future developments. The Las Vegas cityscape was quickly transformed by the fantastical architecture of themed casino hotels—a massive black pyramid, a Disneyesque medieval castle, detailed replicas of the Eiffel Tower and the Statue of Liberty. The skyline now was composed of buildings cribbed from the skylines of other cities: hypertrophied but familiar evocations of place. In other cities, hotels are built in proximity to major attractions; in Las Vegas the hotels *are* the attractions. "Each megahotel is really a nation-state, catering with remarkable precision to a specific socioeconomic group—from Asian high rollers to Midwestern housewives, to NASCAR dads."[12] High-end retailing, with arcades and gallerias of luxury-brand stores, was combined with amenities and spectacle. The concept was "shopertainment."

The degenerate utopia that is the latest iteration of the Las Vegas Strip has brought thousands of upper-middle-class "relos" with their own spending power. Their natural habitat consists of stratified and packaged suburban sprawl. Summerlin, created by the Rouse Group, is the largest and most hyped: "Sparklingly clean, lavishly landscaped and gated, the Summerlin communities offer a range of housing options, from so-called starter homes to million-dollar mansions, and amenities such as golf courses, jogging trails, . . . all intended to suggest the homey innocence of a mythic suburban America. . . . [Summerlin] with its baseball diamonds, parks, libraries, schools, churches, and carefully-zoned shopping districts—promised residents complete insulation from the wanton excesses of Las Vegas."[13] In its promise of inclusion and community, Summerlin both acknowledges and tries to make amends for the fact that the parts of Las Vegas—the casinos, hotels, and themed architecture that make it a tourist destination—do not add up to an integrated community.[14]

Nevertheless, what the relos do not have is a city with a credible civic infrastructure. One hundred years of neoliberal development, with low or non-existent taxes, have inevitably resulted in an "abrogation of civic infrastructure."[15] Greater access to investment capital has nevertheless brought a degree of cultural sophistication in the form of high-end restaurants, art galleries, and Broadway productions. In Las Vegas, gambling pays for culture and cultural institutions become, essentially, institutions of consumption.[16] Steve Wynn introduced his own brand of cultural tourism to his Las Vegas attractions through the purchase and display of what now amounts to over a half a billion dollars worth of art. "Just as the way out of the museum leads through the shop, the exit from the casino is lined with boutiques and museums."[17] Picasso, Manet, Matisse, van Gogh, Cézanne, Modigliani, Warhol—the images appear on billboards on the Strip and the paintings themselves appear in a gallery on the same floor as the slot machines. What Wynn started, Sheldon Adelson, chairman of the board and principal owner of Las Vegas Sands Corporation, took to a new level by creating The Guggenheim Las Vegas, a seven-story hall at the back of the Venetian resort, with a simulacrum of Michelangelo's ceiling in the Sistine Chapel (which is, of course, in Rome, not Venice), designed to house sequential loans of safe-bet exhibitions such as Norman Rockwell and Pop Art. Thus casino transmutes into resort, which transmutes into gallery, which transmutes into consumption. More important, the bigness, bling, and unrestrained vulgarity of Las Vegas has served as a model for many other settings throughout the United States, facilitating and encouraging the competitive consumption of the America's affluent upper-middle classes on an epic scale.

Questions of Taste

While Las Vegas is the undisputed capital of schlock-and-awe urbanism, unrestrained and unashamed vulgarity in taste and behavior can be found throughout metroburbia. Vulgarity, in fact, is by no means a new element of urban and suburban landscapes, nor is it exclusive to American cities. Vulgarity is seemingly a universal human trait, associated especially with the ostentatious consumption of the newly affluent. The nouveau riche Victorians built large and elaborate Italianate, Gothic Revival, Neo-Grecian, and Romanesque villas and mansions that were regarded, by the diamond-hard snobbery of established classes at the time, as monstrous affectations. Now, we see the indulgences of Victorian villas and mansions fondly, as a valued legacy of the *zeitgeist* of an expansive and transitional era. What this points to, of course, is the potentially sensitive issue of how we define vulgarity. At the most general level, vulgarity is conventionally defined as "characteristic or associated with the majority of ordinary people." More specifically, vulgarity is understood as "showing a lack of taste or reasonable moderation"; "ostentatious or excessive in expenditure or display; pretentious." A vulgarian is "someone who is wealthy but tasteless or overly ostentatious."[18] All of these definitions are fine, up to a point, but they are context-dependent. Values and behaviors change over time, and vary from place to place, class to class, and from one affectual or "neotribal" grouping to another. There is also the touchy subject, in a consumer democracy, of taste: how it is defined, and who defines it. This is where Bourdieu's concept of economic, social, cultural, and symbolic capital is especially useful.[19] In terms of contemporary market segments, only the established upper-middle classes—Innovators—have comfortable amounts of symbolic capital (i.e., the command of superior taste) to go along with their economic, social, and cultural capital. Achievers have fairly high levels of economic capital, rather less by way of social capital, and relatively low levels of cultural and symbolic capital, often limited to knowledgeability of established prestige products, services, and behaviors that, while demonstrating success to their peers, may well be seen by others as crass or vulgar.

Questions of taste and the importance of the cultural and symbolic capital associated with people's ensembles of material possessions became much more prominent in America in the democratic utopia of post–World War II affluence. American social historians like Robert Bocock, Lizabeth Cohen, William Leach, and Susan Strasser have argued that this period saw the emergence of a novel complex of social practices and norms that reconfigured popular and democratic American culture around the ideals of individual consumption. And, as Alan Latham has noted, it was a set of social practices and norms that was quickly reconfigured, reimagined, and perfected as a profoundly suburban culture.[20]

As such, it was the apotheosis of the American Dream. The very poor were excluded from this, of course; and most of the very rich opted out, feeling no need to let the rest of the world know about their lifestyle. But for the rest, as Juliet Schor points out in *The Overspent American*, what we acquire and own is tightly bound to our personal identity.

In comparison with other developed countries, American culture has always seemed more materialistic, more disposed toward casual vulgarity and kitsch, less sensitive to aesthetics. In the nineteenth century, French prime minister Georges Clemenceau's celebrated one-liner was that America is the only nation in history to have passed directly from barbarism to decadence without the usual interval of civilization. In 1927, H. L. Mencken famously wrote, "I have seen, I believe, all of the most unlovely towns of the world; they are all to be found in the United States. . . . On certain levels of the American race, indeed, there seems to be a positive libido for the ugly." Describing the towns, suburbs, and houses of the Westmoreland Valley in Pennsylvania, he could not believe that "such masterpieces of horror" could be merely the result of "mere ignorance" or "the obscene humor of the manufacturers" and concluded that the ugliness of the built environment "has been yielded to with an eagerness bordering upon passion."[21] More recently—in 1995—an Italian visitor, Beppe Severgnini, spent a year in the United States and concluded that "the residents of this country are convinced that anything good has got to be over-the-top, in-your-face, and ear-splittingly loud. We might call it large-scale wanton tackiness. . . . The hero figures of this America are Mae West, Liberace, Muhammad Ali, Joan Collins, and Ivana Trump. Larger than life personalities who at first sight, and often at second or third, are beyond comprehension."[22] An updated list might include Donald Trump, Conrad Black, Shaquille O'Neal, and the Beckhams. The conservative commentator David Brooks has characterized the United States as a sort of world "bimbo." America's image to the world, he says, is "what southern California's image is to the rest of America."[23] Today, when foreign observers look at America they see, to paraphrase Brooks, the culture of Disney, McDonald's, MTV, botox, boob jobs, Bart Simpson, and Britney Spears. They see a country obsessed with celebrity; a country that invented Prozac, credit counseling, monster trucks, Cheez Whiz®, and competitive cheerleading; a culture that finds its expression through commercial jingles, vanity license plates, bumper stickers, and T-shirts. They see a consumer culture that has perfected big-box retail, Jewelry Television, the Celebrity Shopping Network, megamalls, and home equity loans.

The New Metroburban Affect

In the formative years of American consumerism, Fordism was the dominant paradigm: mass production for mass consumption. Questions of taste, sophistication, and style were precluded by the standardization of products, from washing machines to tract homes, whose most important attribute was affordability. In keeping with the democratic ethos of the postwar boom, American consumers were more concerned with avoiding below-par experiences than with experiencing the unique or the extraordinary. Standardization and affordability were implicit in the Holiday Inn slogan: "The Best Surprise is No Surprise," and in the brand identities of the likes of McDonald's, Sears, and JC Penney. Meanwhile, American society, the most open in the world in terms of social mobility, had little room for the kind of sophisticated understatement, modesty, or self-restraint that is so highly valued in European and Japanese society. In an open society based on economic competition, understatement and self-denigration could too easily be construed as failure or, worse, dorkiness. So while standardization and affordability were the cornerstones of postwar consumer culture, material success and individual achievement, where and when it occurred, had to be celebrated and signaled loud and clear.

At the risk of invoking a degree of determinism, it should also be pointed out that this critical, formative period of American suburbanization and its associated material culture took place in a very different environmental context from that of Europe or Japan. In the United States, there is an abundance of developable land, with relatively few constraints on building. Upward mobility is easily expressed in sprawling suburbs and ever-larger homes in which to install the standardized but ever-larger range of consumer durables, from washing machines, dishwashers, dryers, and second and third vehicles to ride-on mowers and log splitters. In Europe and Japan, developable space is relatively scarce, planning controls are more stringent, and land and house prices are consequently driven high. Thus when European and Japanese consumers began—a little later than their American counterparts—to enjoy postwar prosperity, living space was still constrained. Standardized consumer durables had to be designed to fit smaller spaces; and as a result there emerged a much stronger emphasis on product design as such. Meanwhile, in the United States the low-density, Fordist suburbs acquired their own loud-and-clear aesthetic. For all practical purposes accessible only by car, the suburban shops and shopping centers of the 1960s and 1970s brought with them acres and acres of parking space. To accommodate this parking space, buildings had to be set back so far from the road that signs became larger and flashier and storefronts and buildings more outlandish in an attempt to catch the motorist's eye.[24] The net result is the classic strip of drive-through banks and fast-food outlets, motels, cow-, chicken-, and donut-shaped buildings, automobile dealerships, gasoline service stations, and mini-marts.

A second important phase in the evolution toward less restraint and more casual vulgarity in Americans' consumption practices came in the so-called "decade of greed," between 1979 and 1989, when the rich and super-rich took conspicuous consumption to new levels. In the 1980s, the top 1 percent of American households increased their annual incomes from an average of about $280,000 to $525,000. The lifestyles of the rich and famous became a source of titillation for popular magazines television programs, and their "visible public excesses" reverberated through American society with dramatic effect. "To compensate for the growing chasm between their lifestyles and those of the rich and famous, the upper-middle classes also began conspicuously acquiring the luxury symbols of the 1980s—buying the high-prestige watches and pens, looking for 'puro lino' labels, and leasing luxury vehicles they often couldn't afford." Lifestyles of the rich and famous thus came to be echoed by those of the affluent and vacuous.

The significance of this mimetic transformation rests in the fact that in the United States, as elsewhere, it is the upper-middle classes, not the upper class or super-rich, that define material success, luxury, and comfort for nearly every class fraction below them. Theirs, as Juliet Schor observes, "is the visible lifestyle to which most Americans aspire."[25] As outlined in chapter 7, the upper middle classes let rip in the 1980s with competitive luxury spending, bolstered by a big tax break from the Reagan administration, by trends in financial markets, by the escalating paychecks of the new economy, and, not least, by easily available credit. Americans now owe more than $750 billion in revolving debt, according to the Federal Reserve: a sixfold increase from two decades ago. Over the last twenty years, the industry has become increasingly indulgent about extending credit and increasingly generous in how much it would let consumers borrow, as long as those customers were willing to pay high fees and risk living in debt. Credit has given so many Americans access to such a wide array of high-end goods that traditional markers of status have lost much of their meaning. In their place has come an increasingly crude calculus of status: the bigger, the newer, the more expensive, the flashier, the better. Lifestyle magazines like *Architectural Digest, Flaunt, GQ, Luxe, Millionaire, More, The Robb Report, Self*, and *Stuff*, along with "reality" television, amplified this conspicuous, competitive, and size-matters consumption: "The more people watch television, the more they think American households have tennis courts, private planes, convertibles, maids, and swimming pools" and that Americans "are millionaires, have had cosmetic surgery, and belong to a private gym."[26] A "horizontal desire" (coveting a neighbor's goods) was replaced by a "vertical desire" (coveting the goods of the affluent, as portrayed in magazines and on television). The net result was that the aspirations of most other Americans were ratcheted sharply upward. Under the influence of the Paradise Spell, "affluenza" became endemic and the American Dream was transmuted into the American Dream Extreme.

In this supercharged materialism, what we lack in taste and sophistication we can make up for in size and ostentation. The cultural hegemony of bigness and bling is transcendent. David Brooks is always entertaining as well as observant on such matters. On bigness:

> At some point in the past decade, the suburbs went quietly berserk. As if under the influence of some bizarre form of radiation, everything got huge. The drinks at 7-Eleven got huge, as did the fry containers at McDonald's. The stores turned into massive, sprawling category-killer megaboxes with their own climatic zones. Suburbia is no longer the land of tick-tacky boxes on a hillside where everything looks the same. It's the land of the gargantuoids.[27]

On bling:

> Highly educated folk don't want to look materialistic or vulgar, but on
> the other hand it would be nice to have an in-house theater with a four-
> teen-foot high-definition projection screen to better appreciate the inter-
> views on Charlie Rose. Eventually these advanced-degree moguls cave
> in and buy the toys they really want: the heated bathroom floors to pro-
> tect their bare feet, the power showers with nozzles every six inches, the
> mud rooms the size of your first apartment, the sixteen-foot refrigera-
> tors with the through-the-door goat cheese and guacamole delivery sys-
> tems, the cathedral ceilings in the master bedroom that seem to be
> compensation for not quite getting to church. Later, when they show off
> to you, they do so in an apologetic manner, as if some other family
> member forced them to make the purchase.[28]

The Moral Landscapes of Contemporary Metroburbia

In a free-enterprise democracy the consumer is king. One may question or
deplore an individual's taste, but individual decisions on consumption are people's own business. It is at the aggregate level that things get interesting. The
overt outcomes of individual decisions on residential location and domestic
spending—patterns of sprawl, land use, infrastructure provision, transportation,
lifestyle, segregation, and so on—are the legitimate focus of politics and public
debate and the traditional focus of public policy and planning. There are, however, less tangible outcomes that warrant attention. Not least of these are the
moral codings of spaces and people that are embedded in our landscapes. The
taken-for-granted familiarity of urban landscapes is a powerful cultural and
political influence: a stealthy, sometimes hallucinogenic normality that both
reflects and reinforces the implicit values that are written in to the built environment. As Pierre Bourdieu has observed, "The most successful ideological effects
are those that have no words, and ask no more than complicitous silence."[29] In a
society dominated by divisions between success and failure, affluence and
poverty, winners and losers, the most powerful moral landscapes of all are
arguably those at the leading edges of the New Metropolis, home to the success-
oriented, affluent "winners" of American society. They are the landscapes of
what I have called "Vulgaria."[30]

Vulgaria
In Vulgaria, nothing succeeds like excess. The outcome of the American
Dream Extreme is a proliferation of landscapes where almost everything is dominated by a presumed reciprocity between size and social superiority. At the
leading edges of the New Metropolis, ostentation and simulation pass for style
and taste, while affluence tends to be confused with cosmopolitanism and urbanity
(Figure 8.1). Vulgaria is characterized by inert and pretentious neighborhoods,

FIGURE 8-1. Plain ostentation. A new home in McLean, Virginia. (Paul L. Knox)

irradiated by bigness and spectacle. Its homes are tract mansions and starter castles of 4,000 square feet and upward, featuring two-story entrance halls, great rooms, three- or four-car garages, huge kitchens, spa-sized bathrooms, his-and-hers room-sized master closets, media rooms, fitness centers, home offices, high-tech security systems, and perhaps even an au pair suite. Vulgaria's exterior residential styling deploys any kind of hybridized neotraditional motif as long as the street frontage is impressive, with high gabled roofs, unusually shaped windows, and architectural features such as turrets, bays, and portes-cochère. The overall effect is an outlandish brashness of contrived spectacle, serial repetition, and over-the-top pretension. It is nouveau riche tackiness on an unprecedented scale.

While dominant at the leading edges of the New Metropolis, Vulgaria is not always an extensive landscape or a coherent zone or sector of the metropolitan mosaic. The landscapes of Vulgaria are cultural kudzu, their bigness, bling, and ostentation filtering down into older suburbs as monster homes replace teardowns and scrapeoffs, smothering greenfield sites as new master-planned developments are themed and packaged, and even creeping into old money neighborhoods. In their study of Bedford, New York, for example, James and Nancy Duncan cite a resident as saying "Bedford used to look like old money. I'm afraid that it has changed for the worse over the past ten years. Everywhere you look, a fancy developer-special has popped up on a four-acre lot. It doesn't look like Bedford any more; it looks like Scarsdale . . . it looks nouveau riche." The sentiment was echoed by the local newspaper, which dismissed new "developer-homes" as "McMansions" and "starter castles."[31]

Vulgaria is so widespread that it has naturalized the neoliberal ideology of competitive consumption and disengagement from notions of social justice and civil society that has been fostered within private, master-planned communities.

Physically designed and tightly regulated through homeowners' associations to provide privacy, autonomy, stability, security, and partition, private, master-planned communities have propagated a kind of moral minimalism in their residents: Bound only by their contracted commitment to lead a private life, most residents have little social contact with neighbors, virtually no social interaction beyond their workplace, and, as a result, few bonds of mutual responsibility. Most are broadly and insouciantly indifferent to issues that go beyond their own property and lifestyle. Although many private master-planned developments derive their enchantment, new urbanist style, from the iconography and rhetoric of intellectuals' utopias of the nineteenth century, they are about as far away from progressive arcadian utopias as it is possible to imagine. Andrew Jackson Downing's promise of a "popular refinement" through suburban design and institutions now seems a quaint idea, as does Olmsted and Vaux's vision of suburbia as an environment that would foster restraint and decorum. Social and political life *has* been entirely reconstructed in suburbia, but not in the way that Patrick Geddes, Peter Kropotkin, Charles Fourier, William Morris, William Ruskin, Clarence Perry, or Frank Lloyd Wright had envisaged. The old ideal, implicit in the original American Dream—of individual economic, social, and residential mobility played out in cities in which a common culture could be forged—has been precluded by the colonization of suburbia into artful but exclusionary fragments of architects' and developers' imaginations. The leading edges of American metropolitan regions have been transformed from democratic utopias and frontiers of social mobility into arenas of bourgeois vulgarity.

Size Matters: Living Large

We have already seen (in chapter 4) how the suburbs have become supersized as consumers have opted for big homes in order to project affluence and status and to accommodate lifestyles that involve a long list of features (Figure 8.2). But it is not just the size of homes and their trappings that have inflated the suburbs. The indulgences of Vulgaria have to be accommodated with an infrastructure of big-box stores, luxury malls, megachurches, and extravagant and exclusive country clubs—often, of course, reached by way of big vehicles. Living large adds up to a material culture dominated by bigness and spectacle, and a sociocultural ecology dominated by competitive spending.

Conspicuous Construction and the Habitus of Vulgaria.   Big developers have been quick to meet the demand for raw size in suburban homes, with the result that the average floor area of new single-family homes in the United States grew by 36 percent between 1985 and 2005. The increased size has made room for larger kitchens, bathrooms and garages, as well as ever-larger ensembles of equipment, appliances, and consumer goods. It has also made room for the rearrangement of the internal organization of domestic space. As Kim Dovey has shown, a broad "genotype" has developed within the market for upscale new suburban homes, characterized by a set of structured relations between four

(a)   (b)

(c)   (d)

FIGURE 8-2. Size matters. Custom homes in The Reserve, a Miller and Smith development in Fairfax County, Virginia. (Paul L. Knox)

primary clusters of space: a formal living zone, an informal living zone, a master suite, and a minor bedroom zone. The deployment and outfitting of these zones is key to establishing, sustaining, and extending the habitus of households in Vulgaria: they are "structuring structures" (see chapter 7), a durable framework that serves to sustain lifestyles in accord with owners' dispositions and allows them to exercise their economic, social, cultural, and symbolic capital.

The formal living zone consists of an entrance way, formal living and dining areas, a stairway, and a den or study. It is typically a highly ritualized space, with high symbolic value and relatively low use value. Its primary function, as Dovey notes, is awe-inspiring symbolic display, "conspicuously framing the pathway of guests from the entry to the new informal zone. The visibility of the formal zone from the point of entry and the entry pathway is key." In up-market homes, dining and formal living areas typically emphasize volume, with high windows and cathedral ceilings. The den or study "generally has double doors opening onto the formal entrance area where it becomes a display of the more intimate pursuits of the owners. When the master bedroom suite adjoins the formal living zone it is also placed on display through double doors for a similar glimpse of intimacy."[32]

The informal living zone incorporates a kitchen with a nook or dining area, plus family room and outdoor living space. Dovey observes that the carefully crafted "informality" of this zone has become the primary setting for social performance. As a result, it now requires fancy design features such as cathedral ceilings and skylights, expensive furnishings, and state-of-the art equipment. "In this formalized informal zone the image of one's informal lifestyle must be carefully monitored and new retreats ('games,' 'recreation' and media') have emerged to cope with the chaos. . . . Sometimes the informal area is called a 'great hall' linking images of status, power, and baronial splendour. It is in this new oversized heart that the dream of home life is increasingly focused." The informal zone typically incorporates an outdoor living area that wraps around the interior space to generate visual links from indoors. But the productive and service aspects of backyards (garbage, washing line, heat pumps, etc.) are excluded from these views. "The backyard is transformed from a place of production into one of consumption as vegetable garden and solar clothes dryer are displaced by swimming pools, electric clothes dryers and designer landscapes."[33]

Master suites consist of a master bedroom and master bathroom and a dressing area, sometimes with the addition of a "retreat" and deck or court. This zone of the genotype caters to the private, more narcisisstic dimensions of consumption, with steam showers, walk-in bathtubs, jet shower panels, infrared saunas, dressing areas with walk-in closets and walls of mirrors, and high-technology beds with super-high threadcount sheets. The minor bedroom zone is often located well away from the master suite. It incorporates children's and guest bedrooms, bathrooms, and recreation areas. The children's bedrooms are typically signified by gender stereotypes—pink girl's rooms with tea-party settings, doll houses, and cushioned window seats; boy's rooms filled with sports images and equipment—and are as much settings for the display of children's material possessions as they are for sleeping or play.

In addition to these zones is the garage, which has evolved into the core of a service zone that often includes laundry, workshop, and storage space in addition to shelter for vehicles. According to the National Association of Home Builders, one in five new homes now have three-car garages, and many of the extra-large garages include utility sinks and other features, with second-floor rooms that can be used for play areas, home offices, work areas, or guest space. The importance of the garage and its accessibility from the street has encouraged "snout house" designs, with a garage protruding in front, shortening the driveway and presenting an array of mundane facades (Figure 8.3).

Lawns and gardens may also be considered part of the genotype, since they, like the homes themselves, are a key component of habitus. The display of high-maintenance yard space demonstrates social capital, which is why landscaping occupies so many pages of covenants, controls, and restrictions in master-planned communities. As with the appearance of the homes themselves, the landscaping is formulaic, with airbrushed lawns, courtesy of powerful chemicals and Hispanic landscaping crews, and borders of shrubs, neatly mulched and trimmed. Though lawns may be quite extensive, the occupants of the houses rarely set foot on them.

Figure 8.4 shows the floor plan of Toll Brothers Hampton model, a close match to the genotype described by Dovey. Buyers can select from a number of variations, using Toll Brothers Design Your Own Home™ online software. The example shown here includes an optional great room, expanded family room,

FIGURE 8-3. A suburban street in Colorado Springs with homes based on a "snout house" design. (Robert E. Lang, The Metropolitan Institute, Virginia Tech)

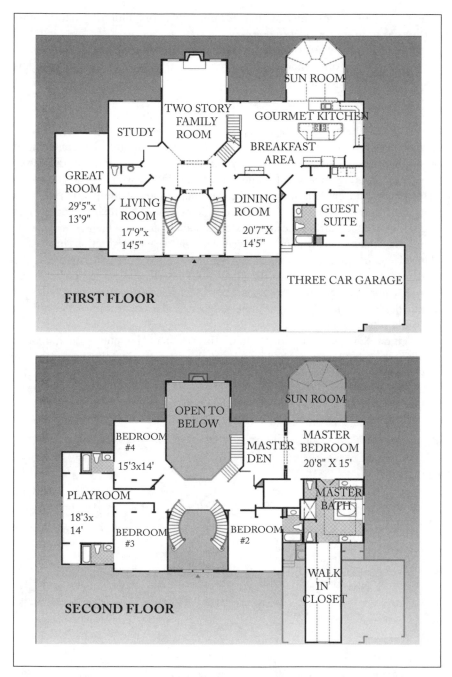

FIGURE 8-4. The Toll Brothers Hampton floor plan. (The floorplans and elevations of Toll Brothers homes are copyrighted. Toll Brothers has enforced and will continue to enforce their federal copyrights to protect the investment of their homebuyers.)

guest suite, sunroom, three-car garage, curved staircase, and a playroom on the second floor. Other options offered on the Hampton floor plan include a butler's pantry, an interior wet bar, a designer fireplace, an additional powder room, a screened-in porch, a solarium, a greenhouse with patio door, a Palladian kitchen, a bonus room, and a fifth bedroom.[34] Design options include multiple choices for windows and doors, interior carpentry, plumbing, air handling and purification systems, appliances, cabinetry, bath fixtures, and so on.

Finally, an extreme but nevertheless telling example of conspicuous construction is the emergence of mini-me play homes: lavish replicas that can include such grown-up amenities as working kitchens, central air, bathrooms, hardwood floors, and media rooms with satellite television. Lilliput, a playhouse manufacturer in Finleyville, Pennsylvania, offers a dozen styles, including Tudors and Victorians, and will also custom-build replicas of parents' homes.[35] Lilliput's playhouses start at about $13,000, with more elaborate models costing as much as $50,000. Denver-based La Petite Maison's playhouses start at $8,000, but top sellers are in the $25,000 range, with prices running as high as $75,000 or more, depending on the add-ons.[36] Some of these mini-me playhouses are so large that they are subject to suburban building ordinances. In Friendswood, Texas, just outside Houston, a two-story, fourteen-foot-tall replica playhouse in the backyard of a Tudor-style home was found by city inspectors to be in violation of several city building codes because of its size, height, and proximity to a neighbor's home. The owners were told to remove it or face as much as $2,000 a day in fines.[37]

Big Boxes and Megastructures.   Bourdieu's concept of habitus exists in different "fields" of life. In every field, there are key signifiers that serve as a frame of reference for checking one another out. Housing is one principal field, but others within which specific class fractions develop and sustain a distinctive habitus include food, art, music, clothes, and even religion. Interestingly, these same fields can also be seen as key to the affective intensities of contemporary urban life emphasized by Thrift. In Vulgaria, they are characterized, like housing, by bigness, spectacle, and bling. Increasingly, they are accommodated in packaged and rebundled developments—integrated megastructures and macrobuildings, business parks, shopping and entertainment complexes, airport complexes, hotel and convention centers, resort complexes, and refurbished heritage and cultural zones—that are in turn characteristic of the fragmenting and polycentric New Metropolis.

Most ubiquitous and characteristic of Vulgaria are what George Ritzer numbers among the "cathedrals of consumption" of contemporary society: big-box chain stores, themed restaurants, superstores, shopping malls, and shopping villages.[38] Big-box stores have become emblematic of the ugly and unrefined character of urban design in the New Metropolis. Favored by retail chains, discount buyer clubs, superstores, and department stores, they are typically windowless structures of 75,000 to 250,000 square feet on one level, with cheap, concrete block construction, easy access by trucks and automobiles, and acres of parking space for customers. Outlet malls and specialty malls usually consist of an integrated series of big boxes, often with lame attempts to disguise their bloated architectural scale through the addition of a veneer of neotraditional trim on their façade. The affluent upper-middle classes of Vulgaria shop at these big-box stores for many of their needs and a few of their wants. Apart from anything else, they have little choice. "In the face of the big boxes' aggressive expansion, local drugstores, stationery stores, clothing stores, and hardware stores have disappeared by the tens of thousands, changing the shape of older suburbs and small towns. . . . Through the 1990s the big boxes 'killed' older malls, chain supermarkets, chain drugstores and small department stores, as well as little markets, pharmacies, and clothing stores."[39]

For most of their wants, however, the upper-middle classes shop at upscale shopping malls and themed and landscaped shopping villages (or else online). These are the retail settings that rely on affect rather than economies of scale to draw in customers, and the affect is, generally, luxury and spectacle.[40] Take, for example, The Mall at Millenia, Orlando, Florida (Figure 8.5), anchored by Neiman Marcus, Bloomingdale's, and Macy's and with stores by Burberry, Chanel, Jimmy Choo, Dior, Gucci, Salvatore Ferragamo, St. John, Tiffany & Co., and Louis Vuitton, and where "breathtaking architecture creates a distinct environment that exceeds all expectations. The latest runway fashions from NY, London, Paris and Milan are broadcast on LED screens atop 35-foot high masts. Amenities such as valet parking, multi-lingual guest services personnel, foreign currency exchange,

FIGURE 8-5. Macy's Court at the Mall at Millenia, Orlando, Florida. (Courtesy of Macy's Court at the Mall at Millenia)

and global shipping at a U.S. Post Office are offered to enhance the shopping experience. . . . The Mall at Millenia offers the finest collection of luxury brands in Orlando. Made for those who surround themselves in luxury, it is an experience that defines elevated living. This superb collection of the world's most desired retailers is unrivaled by any other shopping destination."[41]

As Ritzer notes, the spectacle of the mall has led inexorably to the megamall, "where the objective is to create much larger settings, to make even more spectacular use of space. For example, as of 2003 the Mall of America [in Minneapolis] encompassed more than 500 stores and restaurants on 2.5 million square feet. . . . The colossal size of the Mall of America permits it to house not only a shopping mall but also a theme park (Knott's Camp Snoopy) within its confines. The enchantment lies in the fact that one of our fantastic means of consumption—the theme park—is simply a small part of an even more fantastic means—the mega-mall. In addition to the shopping mall and the theme park, the Mall of America encompasses Underwater World . . . [where] visitors are surrounded by 8,000 sharks, stingrays, and many other types of sea life. Then there is the admission-free Lego Information Center that includes more than thirty Lego constructions, play areas for children, and, need we say, a Lego shop. Catering to adults is the fourth-floor entertainment district that includes, among other things, a 14-screen movie theater, Planet Hollywood, and Hooters."[42]

In the New Metropolis, religion has been branded and commodified in parallel fashion. James Twitchell observes that "in America, the religion business . . . is a marketing free-for-all." Spirituality aside, religion is very big business indeed. Americans spend $4 billion a year on Christian merchandise alone: DVDs, CDs, books, and merchandise from movies like the *Passion of the Christ*.[43] Twitchell goes on to note: "If you want to succeed in the American market, you'd better make church compelling."[44] Megachurches are an outcome of this, deploying size and spectacle to take their place among suburbia's other cathedrals of consumption. The conventional definition of a megachurch is based on Protestant congregations that share several distinctive characteristics: 2,000 or more persons in attendance at weekly worship; a charismatic senior minister; a seven-day-a-week congregational community; and multiple social and outreach activities. In 1970 there were about ten such churches in the United States; this figure grew to 250 by 1990 and by 2005 there were more than 1,200. They have been called the "soul of the exurbs" and it is certainly the case that the most sprawling metropolitan regions—in the South, Southwest, and West—have the most megachurches. The ten largest are Lakewood Church (Houston; average weekly attendance 30,000), Saddleback Valley Community Church (Lake Forest, California; 22,000), Willow Creek Community Church (South Barrington, Illinois; 20,000), Southeast Christian Church (Louisville; 18,757), The Potter's House (Dallas; 18,500), New Birth Missionary Baptist Church (Lithonia, Georgia; 18,000), Calvary Chapel (Fort Lauderdale; 18,000), Crenshaw Christian Center (Los Angeles; 17,000), Calvary Chapel of Costa Mesa (Santa Ana, California; 16,500), and Second Baptist Church (Houston; 16,000).[45]

Lakewood Church is actually a converted sports arena: the Compaq Center, once home to the Houston Rockets. The Compaq Center's twenty luxury suites have been turned into electrical rooms from which Lakewood's twenty-five-person production team operate lighting, curtains, and cameras. Waterfalls book-end the 200-person choir, which is accommodated on a state-of-the-art hydraulic stage. Willow Creek Community Church is a sprawling complex on a landscaped 155-acre site that includes not only two sanctuaries but also a gymnasium that serves as an activity center, a bookstore, a food court, and a cappuccino bar. With no traditional Christian religious symbols—no spires, steeples, bell towers, pointed arches, or crucifixes—on the exterior, it looks more like a performing-arts center, a community college, or a corporate headquarters rather than a place of worship. Many megachurches have found that empty big-box stores can meet the needs of growing congregations. Such properties provide large spaces in high visibility locations that are convenient to worshippers, with built-in park-ing. In 2001, for example, Memorial Drive United Methodist Church in Houston bought a nearby 75,000-square-foot strip center, using some storefronts for its youth and adult programming while maintaining tenants such as TJ Maxx and CVS Pharmacy.

More typical is Radiant, a megachurch in the Phoenix exurb of Surprise. Its new 55,000-square-foot church looks more like an overgrown ski lodge than a place of worship. The foyer includes five fifty-inch plasma-screen televisions, a bookstore, and a cafe with a Starbucks-trained staff making espresso drinks. Krispy Kreme donuts are served at every service, generating an annual Krispy Kreme bill of more than $16,000. For children there are Xboxes—ten of them for fifth and sixth graders alone. The dress code is casual, and there are no crosses, no images of Jesus, nor any other form of religious iconography. "Bibles are optional (all biblical quotations are flashed on huge video screens above the stage). Almost half of each service is given over to live Christian rock with simple, repetitive lyrics in which Jesus is treated like a high-school crush."[46] Jumbotron screens project the lyrics, follow-the-dot style, to happy-clappy songs for a kind of mass karaoke. As Jonathan Mahler observes, megachurches provide a locus for commu-nity in sprawling, decentralized exurbs like Surprise. Megachurch facilities are designed to colonize every aspect of life, with round-the-clock activities that include aerobics classes, social services, fast food, bowling alleys, sports teams, aquatic centers with Christian themes, and multimedia bible classes. As such, they function in parallel fashion to Islamic madrassas, sustaining and extending habitus in the fields of religion, civics, and neighborhood.

Conclusion

Successive phases of enchantment and reenchantment have left contempo-rary suburbia a long way from the seers' and intellectuals' utopias and their vision of moral landscapes that would promote beauty, amenity, civility, restraint, decorum, privacy, status, and community. Only privacy and status have survived as attributes of the degenerate utopias at the leading edges of the New Metropolis. To the extent that community has survived, it is community commodified; the classic idea of community exists only in developers' advertising copy and the rhetoric of new urbanists.

The dominant cultural landscape is one that both reflects and propagates the American Dream Extreme. It is commodified and themed, exclusive and exclusionary, a mosaic of master-planned communities and exurban tract developments. In detail, the styles of domestic architecture range from beltway baronial to tract McMansion; from the eye-wateringly vulgar to the merely trying-too-hard. In general, though, the landscape exhibits what Elizabeth Wilson calls the "uniformity of plenty."[47] The common denominator is a self-conscious affluence that is generally expressed in terms of a formulaic and rather sterile palette, with an iconography based on neotraditional design.

A kind interpretation, offered by some, is that in these large houses people are compensating for the lack of public spaces or civic infrastructure of any significance. But the privatized dioramas that constitute the residential fabric of the New Metropolis are the nurseries of the neoliberalism that places private property rights and individual consumption above public amenity and civic infrastructure. There is an intimate connection between housing and consumption, and the effects are circular and cumulative. Cash-out mortgage refinancing allows home equity to fuel spending, and the more disposable income, the more competitive spending; the more that Vulgaria is infused with affordable luxuries; the more cruelly defining social gradations based on crass materialism; and so on. Meanwhile, spending on real estate in Vulgaria inflates land values more generally, making the old-fashioned version of the American Dream that bit less accessible for lower-middle income households.

In terms of the sociospatial dialectic discussed in chapter 1, the landscapes of Vulgaria are an embodiment of neoliberalism as well as a setting for its maintenance and development. Equally, the same landscapes are an embodiment of consumerism and materialism while acting as an advertisement and encouragement to materialistic lifestyles. They are a reflection of the exclusionary impulses within American society while forming a framework for the perpetuation and intensification of social segregation. Vulgaria must also be the source of intense affect, though the nature of the affect will surely depend on the subjects involved: residents or outsiders, Achievers or Experiencers, rich or poor. As we have seen, landscapes are by no means innocent. In the case of Vulgaria, the imagineered landscapes may seem innocuous enough at first sight. The affect for many will be one of orderliness and prosperity. But the moral codings of buildings, spaces, and the ensembles of people's possessions are vectors of a contagious social condition—"affluenza"—that drives people to live beyond their means. Perhaps most significant of all, the extensiveness of Vulgaria has established bigness, bling, privatism, and social exclusion as normal, taken-for-granted, and implicitly desirable dimensions of American life. Mumford's lament, faced with the incipient sprawl of the 1920s, was "more and more of worse and worse."[48] Today, it seems, we are faced with the prospect of more and more of more and more. With private master-planned developments exempt from most municipal land-use planning regulations, an increasing amount of new residential fabric is likely to be framed within such developments, intensifying the arena of competitive consumption at the leading edges of the New Metropolis, deepening the sociospatial divisions within metropolitan regions as a whole, and seriously impeding any attempt at regional and metropolitan-wide planning.

NOTES

CHAPTER 1 *Introduction*

1. See, for example, Karin Knorr Cetina, "Postsocial Relations: Theorizing Sociality in a Postsocial Environment," in *Handbook of Social Theory*, ed. G. Ritzer and B. Smart (London: Sage, 2001), 520–537.
2. Robert Lang and Paul Knox, "The New Metropolis: Rethinking Megalopolis," *Regional Studies* 42 (2008): in press.
3. Edward Soja, *Postmetropolis: Critical Studies of Cities and Regions* (Oxford: Blackwell, 2000), 239.
4. See, for example, Ulrich Beck, W. Bonss, and C. Lau, "The Theory of Reflexive Modernization," *Theory, Culture, and Society* 20 (2003): 1–33.
5. Simon Graham and S. Marvin, *Splintering Urbanism* (London: Routledge, 2001); Peter Hall and Kathy Pain, *The Polycentric Metropolis* (Sterling, Va.: Earthscan, 2006).
6. The term emerged in Internet and media usage around 2005 to capture an important dimension of the New Metropolis: the intermixing of office employment and high-end retailing with residential settings in suburban and exurban areas, along with established dormitory towns and small cities within a metropolitan area that have acquired many of the amenities of a large city. See, for example, the Merriam-Webster Open Dictionary (http://www3.merriam-webster.com/opendictionary/newword_search.php?word=metro; accessed June 29, 2007).
7. Lifestyle centers are typically between 150,000 and 400,000 square feet, compared with 750,000 square feet and upward for conventional indoor shopping malls.
8. See the web site of the International Council of Shopping Centers, *http://www. icsc.org/* (accessed June 30, 2007).
9. Andrew Blum, "The Mall Goes Undercover," *Slate*, April 6, 2005, http://www.slate.com/id/2116246 (accessed June 30, 2007).
10. Tracy M. Gordon, *Planned Developments in California: Private Communities and Public Life* (San Francisco: Public Policy Institute of California, 2004), 49.
11. *http://maplelawnmd.com* (web site of the Maple Lawn Community, accessed July 1, 2007).
12. Umberto Eco, *The Open Work* (Cambridge, Mass.: Harvard University Press, 1989); see also Peter Bondanella, *Umberto Eco and the Open Text* (Cambridge: Cambridge University Press, 1997).

13. See, for example, Deborah E. Popper, Robert E. Lang, and Frank J. Popper, "From Maps to Myth: The Census, Turner, and the Idea of the Frontier," *Journal of American and Comparative Popular Cultures* 23 (2001): 91–102; Henry Nash Smith, *Virgin Land: The American West as Symbol and Myth* (Cambridge, Mass.: Harvard University Press, 1971); Richard White and Patricia N. Limerick, *The Frontier in American Culture* (Berkeley: University of California Press, 1994).

14. Mitchell Schwarzer, "The Spectacle of Ordinary Building," in *Sprawl and Suburbia*, ed. William S. Saunders (Minneapolis: University of Minnesota Press, 2005), 86.

15. See, for example, Jean Baudrillard, *For a Critique of the Political Economy of the Sign* (St. Louis: Telos, 1981); Jean Baudrillard, *The Consumer Society: Myths and Structures* (London: Sage, 1998); Guy Debord, *The Society of the Spectacle* (New York: Zone Books, 1993); Michel de Certeau, *The Practice of Everyday Life*, trans. S. Randall (Berkeley: University of California Press, 1984); Pierre Bourdieu, *The Field of Cultural Production* (New York: Columbia University Press, 1993); Michel Foucault, "Space, Power, and Knowledge," in *Rethinking Architecture*, ed. N. Leach (London: Routledge, 1997); Roland Barthes, *Mythologies*, trans. A. Lavers (London: Paladin, 1973).

16. Edward Soja, "The Socio-Spatial Dialectic," *Annals of the Association of American Geographers* 70 (1980): 207–225. This concept is further developed in chapter 3 of Edward Soja, *Postmodern Geographies* (London: Verso, 1989). See also Michael Dear and Jennifer Wolch, "How Territory Shapes Social Life," in *The Power of Geography*, ed. M. Dear and J. Wolch (London: Unwin Hyman, 1989).

17. Anthony Giddens, *Central Problems in Social Theory* (London: Macmillan, 1979); Anthony Giddens, *The Constitution of Society: Outline of the Theory of Structuration* (Cambridge: Polity Press, 1984); see also C.G.A. Bryant, and D. Jary, eds., *Giddens's Theory of Structuration: A Critical Appreciation* (London: Routledge, 1991).

18. Dear and Wolch, "How Territory Shapes Social Life," 6.

19. Ray Pahl, "Urban Social Theory and Research," *Environment and Planning A* 1 (1969): 143–153; see also Paul L. Knox and Joe D. Cullen, "Planners as Urban Managers: An Exploration of the Attitudes and Self-Image of Senior British Planners," *Environment and Planning A* 13 (1981): 885–892.

20. Colin Campbell, *The Romantic Ethic and the Spirit of Modern Consumerism* (Oxford: Blackwell, 1987).

21. Robert Bocock, *Consumption* (London: Routledge, 1993); Lizabeth Cohen, *A Consumers' Republic: The Politics of Mass Consumption in Postwar America* (New York: Vintage, 2003); Michel Maffesoli, *The Time of the Tribes: The Decline of Individualism in Mass Society* (London: Sage, 1996); M. Featherstone, *Consumer, Culture and Post-modernism* (London: Sage Publications, 1991).

22. Named after Henry Ford (1863–1947), Fordism is known for large-scale assembly-line production and affordable mass-produced consumer goods. Beginning in the 1920s, it became a widespread regime of capital accumulation, based on the interdependence of mass production and consumption.

23. James Twitchell, *Branded Nation: The Marketing of Megachurch, College, Inc., and Museumworld* (New York: Simon & Schuster, 2004); Mark Paterson, *Consumption and Everyday Life* (London: Routledge, 2006), 199–214.

24. George Ritzer, *Enchanting a Disenchanted World*, 2nd ed. (Thousand Oaks, Calif.: Pine Forge Press, 2005).

25. The concept of the structures of building provision is Michael Ball's, and his original formulations can be consulted in his essay "The Built Environment and the Urban Question," *Environment & Planning D: Society and Space* 4 (1986): 447–464. See also Michael Ball, "The Organization of Property Development Professions," in *Development and Developers*, ed. Simon Guy and John Henneberry (Oxford: Blackwell, 2002), 115–136; Patsy Healy, "Models of the Development Process: A review," *Journal of Property Research* 8 (1992): 219–238; Patsy Healy, "An Institutional Model of the development process," *Journal of Property Research* 9 (1992): 33–44; Patsy Healey and Susan Barrett, "Structure and Agency in Land and Property Development Processes," *Urban Studies* 27 (1990): 89–104.

26. Edward Kaiser and S. Weiss, "Public Policy and the Residential Development Process," *Journal of the American Institute of Planners* 36 (1970): 31.

27. See Paul L. Knox and Linda McCarthy, *Urbanization*, 2nd ed. (Upper Saddle River, N.J.: Prentice Hall, 2005), chapter 11; John Logan and Harvey Molotch, *Urban Fortunes: The Political Economy of Place* (Berkeley: University of California Press, 1987).

28. "Bling" first came into usage to refer to flashy jewelry worn especially as an indication of wealth; more broadly it has come to refer to expensive and ostentatious possessions; see the Merriam Webster Online Dictionary (http://www.merriam-webster.com/dictionary/bling) (accessed July 15, 2007).

29. Sharon Zukin, *Landscapes of Power: From Detroit to Disney World* (Berkeley: University of California Press, 1991), 39.

CHAPTER 2 *Prelude: The Serial Enchantment of Suburbia*

1. There are, however, broad typologies and periodizations of the suburbs that are based on built form, metropolitan form, and urban design: see, for example, Dolores Hayden, *Building Suburbia. Green Fields and Urban Growth, 1820–2000* (New York: Pantheon, 2003); Robert Lang, Jennifer LeFurgy, and Arthur C. Nelson, "The Six Suburban Eras of the United States," *Opolis* 2 (2006): 65–72; and Emily Talen, *New Urbanism and American Planning: The Conflict of Cultures* (New York: Routledge, 2005).

2. David Harvey, *Spaces of Hope* (Berkeley: University of California Press, 2000).

3. David Brooks, *On Paradise Drive* (New York: Simon & Schuster, 2004), 110.

4. John Stilgoe, *Borderlands: Origins of the American Suburb, 1820–1939* (New Haven: Yale University Press, 1990); Kenneth T. Jackson, *Crabgrass Frontier: The Suburbanization of the United States* (New York: Oxford University Press, 1985).

5. The argument is made in Emerson's *Nature* (Boston: James Munroe & Co., 1836). See also J. Woods, " 'Build, Therefore, Your Own World': The New England Village as Settlement Ideal," *Annals of the Association of American Geographers* 81 (1991): 32–50; Henry D. Thoreau, *Walden* (1854; reprint, New York: Holt, Rinehart, and Winston, 1961).

6. Frederick J. Turner, *The Frontier in American History* (New York: Henry Holt, 1920).

7. Leo Marx, *The Machine in the Garden* (New York: Oxford University Press, 1964), 12.

8. Ibid.

9. Hayden, *Building Suburbia*; Dolores Hayden, *The Grand Domestic Revolution: A History of Feminist Designs for American Homes, Neighborhoods, and Cities* (Cambridge, Mass.: MIT Press, 1981).

10. Andrew Jackson Downing, *Cottage Residences, Adapted to North America* (New York: Wiley and Putnam, 1844).

11. James S. Duncan and Nancy G. Duncan, *Landscapes of Privilege: The Politics of the Aesthetic in an American Suburb* (New York: Routledge, 2004), 38.

12. Harvey, *Spaces of Hope*, 164.

13. See, for example, Peter Hall, *Cities of Tomorrow* (Oxford: Blackwell, 1988); K. Kolson, *Big Plans: The Allure and Folly of Urban Design* (Baltimore: Johns Hopkins University Press, 2001); David Pinder, *Visions of the City* (New York: Routledge, 2005); Robert Fishman, *Urban Utopias in the Twentieth Century* (Cambridge, Mass.: MIT Press, 1987); William H. Wilson, *The City Beautiful Movement* (Baltimore: Johns Hopkins University Press, 1989).

14. Robert Fishman, "The American Planning Tradition," in *The American Planning Tradition*, ed. Robert Fishman (Washington, D.C.: Woodrow Wilson Center Press, 2000), 10; see also David Schuyler, *The New American Landscape: The Redefinition of City Form in Nineteenth-Century America* (Baltimore: Johns Hopkins University Press, 1986).

15. Hayden, *Building Suburbia*, chapter 4; Mary Sies, *The Suburban Ideal: A Cultural Strategy for Modern American Living* (Philadelphia: Temple University Press, 2000).

16. Pinder, *Visions of the City*.

17. Robert Fishman, *Urban Utopias in the Twentieth Century*; Hall, *Cities of Tomorrow*, 88–91.

18. Hall, *Cities of Tomorrow*, 97.

19. Ibid.

20. John Nolen, *New Towns for Old: Achievements in Civic Improvement in Some American Small Towns and Neighborhoods* (1927; reprint, Amherst: University of Massachusetts Press, 2005).

21. Hall, *Cities of Tomorrow*, 137.

22. G. Woodcock, *Anarchism: A History of Liberation Ideas and Movements* (New York: Meridian Books, 1962).

23. See, for example, Patrick Geddes, *Cities in Evolution* (London: Williams and Norgate, 1915).

24. Lewis Mumford, *The Culture of Cities* (London: Secker and Warburg, 1938); Lewis Mumford, *The Story of Utopias* (London: Harrap, 1923).

25. Wilson, *City Beautiful Movement*; John Chambers, *The Tyranny of Change: America in the Progressive Era, 1890–1920* (New Brunswick, N.J.: Rutgers University Press, 1992).

26. Manfredo Tafuri and Francesco Co, *Modern Architecture* (New York: Rizzoli, 1986), 40.

27. Robert Fishman, *Bourgeois Utopias: The Rise and Fall of Suburbia* (New York: Basic Books, 1987), x.

28. Balloon framing utilizes long continuous framing members that run from sill to eave line with intermediate floor structures nailed to them. Instead of using heavy beams and corner posts held together by mortise and tenon joints, balloon framing uses standardized, machine-cut 2 × 4-inch studs nailed together by inexpensive, machine-cut nails. By spreading the stress over a large number of light boards, the balloon frame had strength and stability far beyond its insubstantial appearance. Using standardized components and requiring only semi-skilled labor, balloon framing was about 40 percent less expensive than traditional methods.

29. The first successful streetcar system was established by Frank Sprague, who had left Thomas Edison to form his own company and experiment with electrically powered omnibus systems. Drawing on the experience of numerous attempts to solve the urban transportation system—including electrified trolley systems in Cleveland, Baltimore, East Orange (New Jersey), Montgomery (Alabama), and, in particular, the Siemens system developed in Germany—Sprague perfected an electrically driven version of the horsecar, powered by way of overhead cables, that opened for business in Richmond, Virginia, in the spring of 1888. The innovation was adopted the following year by Henry Whitney, who had established in Boston an integrated horsecar system that covered the entire city. By electrifying the system, Whitney could extend the radial lines, since streetcars could travel much faster than horsecars. In Boston, the total length of track in 1887 had been just over 200 miles; by 1904 it was almost 450 miles. More than 200 other cities adopted Sprague's system within five years of his success at Richmond.

30. Fishman, "American Planning Tradition," 12; see also Fishman, *Bourgeois Utopias.*

31. Don Mitchell, *Cultural Geography: A Critical Introduction* (Oxford: Blackwell, 2000), 129.

32. The philosophy of classical liberalism was central to the founding documents of the United States—the Federalist Papers, the Constitution, and the Declaration of Independence—having been formally articulated in the works of David Hume, Jeremy Bentham, and Adam Smith.

33. The link between transit and real estate development was particularly strong in West Coast suburban developments. In northern California, F. M. Smith bought and consolidated the trolley lines in San Francisco's East Bay, and purchased 13,000 acres of land for development in the Oakland and Berkeley areas in the 1900s. In Southern California, Henry Huntington, a founder of the Southern Pacific railroad, developed the Pacific Electric Interurban Transit Company in the Los Angeles area. He bought up the land along his routes and laid out suburban developments, at the same time avoiding competitor's land holdings unless he was made a partner. Among Huntington's partners was Harry Chandler, the largest developer in Los Angeles. In the first decade of the twentieth century, Chandler bought 47,500 acres in the San Fernando Valley, an area about the size of the city of Baltimore, and the Pacific Electric extended their lines into the valley. A $25 million water project, paid for by the city of Los Angeles, supplied the development with water after a vigorous campaign for the water project, led by Chandler's father-in-law, Harrington Grey Otis, publisher of the *Los Angeles Times*. Later Chandler bought the 300,000 acre (468 square miles) Tejon Ranch in Los Angeles and Kern counties, land that is still controlled by Chandler interests. See Mike Davis, *City of Quartz: Excavating the Future in Los Angeles* (London: Verso, 1990).

34. Sam Bass Warner Jr., *Streetcar Suburbs* (Cambridge, Mass.: Harvard University Press, 1962).

35. Hayden, *Building Suburbia*, 88.

36. Robert Fogelsong, *Bourgeois Nightmares: Suburbia 1870–1930* (New Haven: Yale University Press, 2005).

37. Marc A. Weiss, *The Rise of the Community Builders: The American Real Estate Industry and Urban Planning* (New York: Columbia University Press, 1987), 3.

38. Matthew Edel, Elliot D. Sclar, and Daniel Luria, *Shaky Palaces: Homeownership and Social Mobility in Boston's Suburbanization* (New York: Columbia University Press, 1984), 1.

39. James T. Adams, *The Epic of America* (New York: Blue Ribbon Books, 1931).
40. Jeffrey M. Hornstein, *A Nation of Realtors: A Cultural History of the Twentieth-Century American Middle Class* (Durham, N.C.: Duke University Press, 2005), 7.
41. The label *realtor* was trademarked in order to differentiate members of the local, state, and national association, who voluntarily subscribe to the Code of Ethics of the organization, from unregistered and, often, unscrupulous brokers of real estate.
42. Hornstein, *Nation of Realtors*, 121.
43. Ibid., 155.
44. Fishman, *Urban Utopias*.
45. Overaccumulation results from episodes of severe mismatch between supply and demand within capitalist economies and results in idle productive capacity, excess inventories, gluts of commodities, surplus money capital, and high levels of unemployment. See Anthony J. Badger, *The New Deal: The Depression Years 1933–1940* (Chicago: Ivan R. Dee, 2002); and David M. P. Freund, "Marketing the Free Market: State Intervention and the Politics of Prosperity in Metropolitan America," in *The New Suburban History*, ed. Kevin M. Kruse and Thomas J. Sugrue (Chicago: University of Chicago Press, 2006), 11–32.
46. See George C. Galster and E. Godfrey, "Racial Steering by Real Estate Agents in the U.S. in 2000," *Journal of the American Planning Association* 71 (2005): 251–265; Arthur C. Nelson and Casey J. Dawkins, "Changing Housing Preferences with Implications for Prices and Planning Policy," paper presented at the 47th Annual Association of Collegiate Schools of Planning Conference, Fort Worth, Texas, November 9–12, 2006.
47. Barry Checkoway, "Large Builders, Federal Housing Programs, and Postwar Suburbanization," *International Journal of Urban and Regional Research* 4 (1980): 21–45; David Harvey, *The Urbanization of Capital: Studies in the History and Theory of Capitalist Urbanization* (Baltimore: Johns Hopkins University Press, 1985); Robert Lake, "Spatial Fix 2: The Sequel," *Urban Geography* 16 (1995): 189–191.
48. See W. M. Rohe and H. L. Watson, eds., *Chasing the American Dream: New Perspectives on Affordable Homeownership* (Ithaca, N.Y.: Cornell University Press, 2007).
49. Peter J. Hugill, "Good Roads and the Automobile in the United States, 1880–1929," *Geographical Review* 72 (1982): 327–349.
50. James J. Flink, *The Automobile Age* (Cambridge, Mass.: MIT Press, 1998), 371.
51. Harvey, *Urbanization of Capital*; David Harvey, *The Condition of Postmodernity* (Oxford: Blackwell, 1989).
52. Lizabeth Cohen, *A Consumers' Republic: The Politics of Mass Consumption in Postwar America* (New York: Vintage, 2003).
53. Hayden, *Building Suburbia*, 148.
54. Robert Beauregard, *When America Became Suburban* (Minneapolis: University of Minnesota Press, 2006), 6.
55. Brooks, *On Paradise Drive*, 268.
56. Colin Campbell, *The Romantic Ethic and the Spirit of Modern Consumerism* (Oxford: Blackwell, 1989).
57. John Archer, *Architecture and Suburbia: From English Villa to American Dream House, 1690–2000* (Minneapolis: University of Minnesota Press, 2005), 332.
58. Mark Gottdiener, *Planned Sprawl: Public and Private Interests in Suburbia* (Beverly Hills, Calif.: Sage, 1977).

59. Jason Hackworth, *The Neoliberal City: Governance, Ideology, and Development in American Urbanism* (Ithaca, N.Y.: Cornell University Press, 2007).
60. Lewis Mumford, "Regions—To Live In," *Survey* 54 (1925): 151.
61. John Keats, *The Crack in the Picture Window* (Boston: Houghton Mifflin, 1957), 1. Quoted in Alex Marshall, *How Cities Work: Suburbs, Sprawl, and the Roads Not Taken* (Austin: University of Texas Press, 2003), 100–101.
62. The editors of *Fortune* magazine, *The Exploding Metropolis* (Garden City, N.Y.: Doubleday, 1958); Lewis Mumford, *The City in History: Its Origins, Its Transformations, and Its Prospects* (New York: Harcourt, Brace & World, 1961), 486.
63. Mumford, *City in History*, 619.
64. Jane Jacobs, *The Death and Life of Great American Cities* (New York: Random House, 1961).
65. "Little Boxes," words and music by Malvina Reynolds, Schroeder Music Co., 1962; "Big Yellow Taxi," written and originally performed by Joni Mitchell in 1970; "My City Was Gone," written by Chrissie Hynde and originally performed by the Pretenders, 1984.
66. See, for example, Real Estate Research Corporation, *The Costs of Sprawl: Environmental and Economic Costs of Alternative Residential Development Patterns at the Urban Fringe*, vol. 1, *Detailed Cost Analysis*; vol. 2, *Literature Review*; vol. 3, *Executive Summary* (Washington, D.C.: U.S. Government Printing Office, 1974); Peter Newman and Jeffrey Kenworthy, *Cities and Automobile Dependence* (Aldershot, Eng.: Gower Publishing, 1989); Robert Burchell et al., *Costs of Sprawl Revisited* (Washington, D.C.: National Academies Press, 1998); Robert Burchell et al., *Costs of Sprawl–2000* (Washington, D.C.: Transportation Research Board, National Research Council, 2002); Michael Sorkin, ed., *Variations on a Theme Park* (New York: Hill and Wang, 1992); Saunders, *Sprawl and Suburbia* (Minneapolis: University of Minnesota Press, 2005); Douglas Morris, *It's a Sprawl World After All: The Human Cost of Unplanned Growth* (Gabriola Island, B.C.., Canada: New Society Publishers, 2005).
67. Jon Teaford, *The Metropolitan Revolution* (New York: Columbia University Press, 2006), 248.
68. Robert Bruegmann, *Sprawl: A Compact History* (Chicago: University of Chicago Press, 2005); Deyan Sudjic, *The 100-Mile City* (London: Andre Deutsch, 1991); Joel Kotkin, "Suburbia: Homeland of the American Future," *Next American City* 11 (August 2006): 19–22, *http://www.americancity.org/article.php?id_article=183* (accessed June 16, 2007). See also Edward Banfield, *The Unheavenly City: The Nature and the Future of Our Urban Crisis* (Boston: Little, Brown, 1968); Bernard Frieden, *The Environmental Protection Hustle* (Cambridge, Mass.: MIT Press, 1979); Peter Gordon and Harry Richardson, *The Case for Suburban Development* (Los Angeles: School of Urban and Regional Planning, UCLA, 1995); Wendell Cox, *War on the Dream: How Anti-Sprawl Policy Threatens the Quality of Life* (New York: Universe, 2006).
69. Alex Krieger, "The Costs—and Benefits?—of Sprawl," in *Sprawl and Suburbia*, ed. Saunders (Minneapolis: University of Minnesota Press, 2005), 53.
70. M. Stein, *The Eclipse of Community* (New York: Harper & Row, 1960); David Riesman, *The Lonely Crowd* (New Haven: Yale University Press, 1950); Herbert Gans, *The Levittowners: Ways of Life and Politics in a New Suburban Community* (New York: Vintage, 1967); R. S. Lynd and H. M. Lynd, *Middletown* (New York: Harcourt, Brace Jovanovich, 1956); William Whyte, *The Organization Man*

(New York: Doubleday, 1956); Bennett Berger, *Working Class Suburb* (Berkeley: University of California Press, 1960).

71. Robert Putnam, *Bowling Alone: The Collapse and Revival of American Community* (New York: Simon & Schuster, 2000); David Brooks, "Patio Man and the Sprawl People," *Weekly Standard*, 7:46 (2002); *http://www.weeklystandard.com/Content/Public/Articles/000/000/001/531wlvng.asp*, (accessed July 16, 2007).

72. James Kunstler, *The Geography of Nowhere: The Rise and Decline of America's Man-Made Landscape* (New York: Simon & Schuster, 1993); Edward Relph, *Place and Placelessness* (London: Pion, 1976); Rem Koolhaas et al., *Mutations: Harvard Project on the City* (Barcelona: ACTAR, 2000), 525; Witold Rybcznski, *City Life: Urban Expectations in a New World* (New York: Scribner, 1995), 196.

73. Martin Heidegger, *Poetry, Language, Thought* (New York: Harper & Row, 1971); Raymond Williams, *The City and the Country* (London: Chatto and Windus, 1973).

74. Note that extreme commuters may not be heading into work five days a week and that technology has allowed them to partly telecommute and thus free them to live great distances from work.

75. ABC News/*Time* magazine/*Washington Post* poll, "Gridlock Nation: America's Traffic Toll," 2005: http://abcnews.go.com/ThisWeek/TheList/story?id=495692 (accessed June 16, 2007; U.S. Bureau of the Census, American Community Survey, 2005: *http://www.census.gov/acs/www* (accessed June 16, 2007; Richard Morin and Steven Ginsberg, "Painful Commutes Don't Stop Drivers," *Washington Post*, February 13, 2005, A01; Steven Ginsberg, "D.C. Area Traffic Heavier, Costlier," *Washington Post*, September 8, 2004, A01.

76. Gregory D. Squires, "Urban Sprawl and the Uneven Development of Metropolitan America," in *Urban Sprawl. Causes, Consequences, and Policy Responses*, ed. Gregory D. Squires (Washington, D.C.: Urban Institute Press, 2002), 1.

77. William Lucy and David Phillips, *Tomorrow's Cities, Tomorrow's Suburbs* (Chicago: Planners Press, 2006), 272.

78. Donna Gaines, *Teenage Wasteland: Suburbia's Deadend Kids* (New York: Pantheon, 1991).

79. Brooke A. Masters and Michael D. Shear, "The Vulnerable Suburbs," *Washington Post*, April 5, 1998, A1.

CHAPTER 3   *Metroburbia and the Anatomy of the New Metropolis*

1. Eric Swyngedouw, "Exit 'Post'—The Making of 'Glocal' Urban Modernities," in *Future City*, ed. S. Read, J. Rosemann, and J. van Eldijk (London: Spon Press, 2005), 126.

2. The U.S. census now has a poly-nuclear "principal city" category that has lifted select municipalities to the status of big cities.

3. On urban realms, see James E. Vance Jr., *This Scene of Man: The Role and Structure of the City in the Geography of Western Civilization* (New York: Harper's College Press, 1977); on megalopolis, see Jean Gottmann, *Megalopolis: The Urbanized Northeastern Seaboard of the United States* (New York: Twentieth-Century Fund, 1961), and T. J. Vicino, B. Hanlon, and J. Short, "Megalopolis 50 Years On: The Transformation of a City Region," *International Journal of Urban and Regional Research* 31 (2007): 344–367; on the galactic metropolis, see Peirce F. Lewis,

"The Galactic Metropolis," in *Beyond the Urban Fringe*, ed. R. H. Pratt and G. Macinko (Minneapolis: University of Minnesota Press, 1983).

4. Michael Dear, "Comparative Urbanism," *Urban Geography* 26 (2005): 248; see also Jon Teaford, *The Metropolitan Revolution* (New York: Columbia University Press, 2006).

5. Edward Soja, *Postmetropolis. Critical Studies of Cities and Regions* (Oxford: Blackwell, 2000); Ulrich Beck, W. Bonss, and C. Lau, "The Theory of Reflexive Modernization," *Theory, Culture, and Society* 20 (2003): 1–33.

6. Robert Lang and Paul Knox, "The New Metropolis: Rethinking Megalopolis," *Regional Studies* 42 (2008): in press.

7. Beck, Bonss, and Lau, "The Theory of Reflexive Modernization."

8. Simon Graham and S. Marvin, *Splintering Urbanism* (London: Routledge, 2001).

9. Peter T. Kilborn, "The Five-Bedroom, Six-Figure Rootless Life," *New York Times*, June 1, 2005.

10. Jackie Cutsinger and George Galster, "There Is No Sprawl Syndrome: A New Typology of Metropolitan Land Use Patterns," *Urban Geography* 27 (2006): 228–252. Sprawl is an elusive concept: see George Galster, Royce Hanson, et al., "Wrestling Sprawl to the Ground—Defining and Measuring an Elusive Concept," in *Rates, Trends, Causes, and Consequences of Urban Land-Use Change in the United States*, ed. William Acevedo, Janis L. Taylor et al., Professional Paper 1726 (Reston, Va.: U.S. Geological Service, 2006), 55–58.

11. Robert Fishman, "Beyond Sprawl," in *Sprawl and Suburbia*, ed. William S. Saunders (Minneapolis: University of Minnesota Press, 2005), xiv.

12. Defined by Joel Garreau in *Edge City: Life on the New Frontier* (New York: Doubleday, 1991) as consisting of at least 5 million square feet of office space and 600,000 square feet of retail space.

13. Peter Hall, "Global City-Regions in the 21st Century," in *Global City-Regions: Trends, Theory, Policy*, ed. A. J. Scott (New York: Oxford University Press, 2001), 59–77.

14. Debbie L. Sklar, "The Next Capital of Cool," *Irvine World News*, February 20, 2003.

15. Robert E. Lang and J. S. Hall, *The Sun Corridor: Planning Arizona's Megapolitan Area* (Tempe, Ariz.: Morrison Institute of Public Policy, 2006).

16. Robert E. Lang, *Edgeless Cities: Exploring the Elusive Metropolis* (Washington, D.C.: Brookings Institution Press, 2003); Robert E. Lang, Thomas Sanchez, and Jennifer LeFurgy, *Beyond Edgeless Cities: A New Classification System for Suburban Business Districts* (Washington, D.C.: The National Association of Realtors, 2006).

17. Michael J. McCarthy, "Main Street America Gets a New Moniker," *Wall Street Journal*, August 23, 2004.

18. Robert E. Lang and Jennifer LeFurgy, *Boomburbs: the Rise of America's Accidental Cities* (Washington, D.C.: Brookings Institution Press, 2007).

19. Carl Abbot, *The Metropolitan Frontier: Cities in the Modern American West* (Tucson: University of Arizona Press, 1993).

20. David Rusk, *Cities Without Suburbs* (Washington, D.C.: Woodrow Wilson Center Press, 1995).

21. William Fulton, *The Reluctant Metropolis: The Politics of Urban Growth in Los Angeles* (Berkeley, Calif.: Solano Press Books, 1997).

22. Bruce Katz, "Welcome to the 'Exit Ramp' Economy," *Boston Globe*, May 13, 2001, A19.

23. Lang and Le Furgy, *Boomburbs*, p. 8.

24. Anthony Flint, *This Land; The Battle over Sprawl and the Future of America* (Baltimore: Johns Hopkins University Press, 2006), 13.

25. Auguste Spectorsky, *The Exurbanites* (Philadelphia: Lippincott, 1955).

26. Paul C. Sutton, "A Scale-Adjusted Measure of Urban Sprawl Using Nighttime Satellite Imagery," *Remote Sensing of Environment* 86 (2003): 353.

27. David M. Theobald, "Landscape Patterns of Exurban Growth in the USA from 1980 to 2020," *Ecology and Society* 10 (2005): 32, available at http://www.ecologyandsociety.org/v0110/iss1/art32/. See also David M. Theobald, "Land-Use Dynamics beyond the American Urban Fringe," *The Geographical Review* 91 (2001): 544–564; Marcy Burchfield, Henry G. Overman, Diego Puga, and Matthew A. Turner, "Causes of Sprawl: A Portrait from Space," *Quarterly Journal of Economics* (2006): 587–633; Arthur C. Nelson, "Characterizing Exurbia," *Journal of Planning Literature* (1992): 350–368.

28. Alan Berube, Audrey Singer, Jill H. Wilson, and William H. Frey, *Finding Exurbia: America's Fast-Growing Communities at the Metropolitan Fringe* (Washington, D.C.: Brookings Institution Press, 2006).

29. These restrictions were overturned by the Virginia Supreme Court in 2005; see chapter 6.

30. Peter Whoriskey, "Washington's Road to Outward Growth," *Washington Post*, August 9, 2004, A01.

31. Center for Housing Policy, *A Heavy Load: The Combined Housing and Transportation Burdens of Working Families* (Washington, D.C.: Center for Housing Policy, 2006).

32. Stuart Leavenworth, "Atlanta Offers Glimpse of Triangle to Come," *Charlotte News and Observer*, July 23, 1997, quoted in Alex Marshall, *How Cities Work: Suburbs, Sprawl, and the Roads Not Taken* (Austin: University of Texas Press, 2000), 45.

33. Lang and Knox, "The New Metropolis."

34. Robert E. Lang and Arthur C. Nelson, "America 2040: The Rise of the Megapolitans," *Planning* (January 2007): 5–12.

35. Ibid., 6.

36. Robert E. Lang and Arthur C. Nelson, *Beyond Metroplex: Examining Commuter Patterns at the Megapolitan Scale* (Cambridge, Mass.: Lincoln Institute for Land Policy, 2007).

37. William H. Lucy and David L. Phillips, *Tomorrow's Cities, Tomorrow's Suburbs* (Chicago: Planners Press, 2006). See also T. Swanstrom, C. Casey, R. Flack, and P. Dreier, *Pulling Apart: Economic Segregation among Suburbs and Central Cities in Major Metropolitan Areas* (Washington, D.C.: Brookings Institution Metropolitan Policy Program, 2004); and William H. Frey, "Melting Pot Suburbs: A Study of Suburban Diversity," in *Redefining Urban and Suburban America: Evidence from Census 2000*, vol. 1, ed. B. Katz and R. Lang (Washington, D.C.: Brookings Institution Press, 2003), 155–180.

38. William H. Frey and Alan Berube, "City Families and Suburban Singles: An Emerging Household Story," in *Redefining Urban and Suburban America: Evidence from Census 2000*, vol. 1, ed. Bruce Katz and Robert E. Lang (Washington, D.C.: Brookings Institution Press, 2003), 257–290.

39. Lang and LeFurgy, *Boomburbs*.
40. Robert Lang and Edward Blakely, "In Search of the Real OC: Exploring the State of American Suburbs," *The Next American City* (August 2006): 16–18.
41. Robert Frank, *Richistan: A Journey Through the American Wealth Boom and the Lives of the New Rich* (New York: Crown, 2007).
42. Myron Orfield, *American Metropolitics* (Washington, D.C.: Brookings Institution Press, 2002).
43. U.S. Department of Housing and Urban Development, *The State of the Cities* (Washington, D.C.: U.S. Government Printing Office, 1999).
44. Lucy and Phillips, *Tomorrow's Cities*, 123.
45. Sandra Tsing Loh, "Kiddie Class Struggle," *Atlantic Monthly* 295 (June 2005): 114, 116.
46. Barbara Ehrenreich, *Fear of Falling* (New York: Pantheon, 1989).
47. Ulrich Beck, *Risk Society: Towards a New Modernity* (London: Sage, 1992).
48. Richard Sennett, *The Conscience of the Eye: The Design and Social Life of Cities* (New York: Knopf, 1990), xii.
49. Robert Reich, "Secession of the Successful," *New York Times Magazine*, Jan. 20, 1991, 17; see also Robert Reich, *The Work of Nations: Preparing Ourselves for 21st-Century Capitalism* (New York: Knopf, 1991).
50. C. Tiebout, "A Pure Theory of Local Public Expenditure," *Journal of Political Economy* 64 (1956): 416–424; E. Ostrom, "Metropolitan Reform: Propositions Derived from Two Traditions," *Social Science Quarterly* 53 (1972): 474–493.
51. Tom Hogen-Esch, "Urban Secession and the Politics of Growth: The Case of Los Angeles," *Urban Affairs Review* 36 (2001): 785.
52. Ontological security is "the confidence that most human beings have in the continuity of their self-identity and the constancy of the surrounding social and material environments of action." See Anthony Giddens, *The Consequences of Modernity* (Stanford, Calif.: Stanford University Press, 1990), 92.
53. Dejan Sudjic, *The Edifice Complex* (New York: Penguin, 2005).
54. Loh, "Kiddie Class," 113.
55. See *http://www.claritas.com/claritas/Default.jsp?ci=3&si=4&pn=prizmne* (accessed February 19, 2007).
56. Patrick Simmons and Robert E. Lang, "The Urban Turnaround," in *Redefining Urban and Suburban America*, ed. Bruce Katz and Rob Lang (Washington, D.C.: Brookings Institution Press, 2003), 51–62.
57. Since 2000, the high-end residential market in the Washington metro region, along with the suburban residential market in general, has been supercharged. In the wake of the 9/11 attacks in 2001, the federal government created the Department of Homeland Security, the largest new agency in half a century; the Defense Department's budget ballooned due to the wars in Iraq and Afghanistan; and private-sector defense and security companies garnered billions of dollars in government contracts and went on hiring sprees. Between 2000 and 2005, about 287,000 jobs were created in the region, about half of them in private-sector firms doing government work. In the same period, median home prices rose by 32 percent. Then, in 2005, the Department of Defense announced a program of Base Realignment and Closure (BRAC) that included the strategic relocation of thousands of jobs away from vulnerable, congested, and expensive locations inside the Beltway to military land in the suburbs.

CHAPTER 4   *Developers' Utopias*

1. Entitlements are secured legal permissions from regulatory agencies, typically in the form of permits, but sometimes in the form of rezoning.
2. Fee simple sales give unqualified ownership and power of disposition to the buyer.
3. Marc A. Weiss, *The Rise of the Community Builders. The American Real Estate Industry and Urban Land Planning* (New York: Columbia University Press, 1987), 39.
4. Ibid.
5. Key pieces of environmental legislation include the 1966 National Historic Preservation Act, the 1968 National Flood Insurance Act, the 1970 National Environmental Policy Act, the 1972 Clean Water Act, the 1973 Endangered Species Act, the 1980 Comprehensive Environmental Response, Compensation, and Liability Act (the "Superfund"), and the 1990 Water Resources Development Act.
6. Single-family home starts dipped from 1,179,400 in 1985 to 840,400 in 1991 before rebounding to 1,715,800 in 2005.
7. These data are from BuilderOnline (*http://www.builderonline.com/*) (accessed June 16, 2007).
8. Jon Gertner, "Chasing Ground," *New York Times Magazine*, October 16, 2005, *http://www.nytimes.com/2005/10/16/magazine* (accessed November 15, 2006).
9. Michele Mariani, "Consistent Performers," *Builder Magazine*, May 1, 2006, *http://www.builderonline.com* (accessed November 16, 2006).
10. Gertner, "Chasing Ground."
11. Bob Fernandez, "Cashing In on Housing Boom," *Philadelphia Inquirer*, April 16, 2006, http://www.philly.com/mld/philly/business/14351168.htm (accessed April 25, 2006).
12. *Notable Corporate Chronologies*, online ed. (Thomson Gale, 2005). Reproduced in Business and Company Resource Center (Farmington Hills, Mich.: Gale Group, 2005).
13. Kathleen Kerwin, "A New Blueprint at Pulte Homes," *Business Week*, October 3, 2005, 76.
14. Harvey Molotch, "The City as a Growth Machine: Toward a Political Economy of Place," *American Journal of Sociology* 82 (1976): 309–310.
15. See, for example, John R. Logan and Harvey Molotch, *Urban Fortunes: The Political Economy of Place* (Berkeley: University of California Press, 1987); and Andrew Jonas and David Wilson, eds., *The Urban Growth Machine: Critical Perspectives Two Decades Later* (Albany: State University of New York Press, 1999).
16. Quoted in Dolores Hayden, *Building Suburbia. Green Fields and Urban Growth, 1820–2000* (New York: Pantheon Books, 2003), 158.
17. Ernst Lamothe Jr., "Firm Apologizes for Ad Criticizing Chili Board," *Rochester (N.Y.) Democrat and Chronicle*, October 10, 2006, http://www.democratandchronicle.com/apps/pbcs.dll/article?AID=/20061010/NEWS01/610100355 (accessed November 17, 2006).
18. William Frey, "Census 2000," *Urban Land* (May 2002): 75–80.
19. Kristen Gerencher, "Generation X May Boost Sagging Real-Estate Market," *Wall Street Journal Guide to Property*, http://www.realestatejournal.com (accessed November 20, 2006).
20. Arthur C. Nelson and Robert E. Lang, "The Next 100 Million," *Planning* (2007): 2–4.

21. Adrienne Schmitz, Pam Engebretson, et al. *The New Shape of Suburbia: Trends in Residential Development* (Washington, D.C.: Urban Land Institute, 2003).

22. Victoria R. Wilbur, *ULI Forum Report* (Washington, D.C.: Urban Land Institute, 2002), 2.

23. U.S. Census Bureau, *American Community Survey 2005, http://factfinder.census.gov/ servlet/DatasetMainPageServlet?_program=ACS&_lang=en&_ts=133802848315* (accessed June 16, 2007).

24. Quoted in Wilbur, *ULI Forum Report* 3.

25. National Association of Homebuilders Economics Group (NAHB), *What 20th Century Home Buyers Want* (Washington, D.C.: National Association of Home Builders, 2001); American LIVES, *Community Preferences: What the Buyers Really Want in Design, Features, and Amenities* (Oakland, Calif.: American LIVES, 1999); *Better Homes and Gardens*, "American Homeowners' Wish List" http:// www.bhg.com/bhg/story.jhtml?storyid=/templatedata/bhg/story/data/BHBL_Survey_ 01112005.xml (accessed June 16, 2007).

26. Rick Lyman, "Living Large, by Design, in the Middle of Nowhere," *New York Times*, August 15, 2005, http://www.nytimes.com/2005/08/15/national/15exurb.html?ex= 1281758400&en=1c3dc1b7a4bd8b9e&ei=5088&partner=rssnyt&emc=rss (accessed June 16, 2007).

27. David Brooks, *On Paradise Drive* (New York: Simon & Schuster, 2004).

28. NAHB, *What 20th Century Home Buyers Want.*

29. Calculated as base hits divided by times at-bat. A base hit is a play in which the batter hits the ball in fair territory and reaches at least first base before being thrown out. Walks, hit-by-pitch, fielding errors, and sacrifice hits are not counted as official at-bats.

30. Lyman, "Living Large."

31. Stephanie McCrummen, "Redefining Property Values," *Washington Post*, April 16, 2006, A01.

32. Dowell Myers and Elizabeth Gearin, "Current Preferences and Future Demand for Denser Residential Environments," *Housing Policy Debate* 12 (2001): 633–659.

33. Anthony Flint, *This Land: The Battle Over Sprawl and the Future of America* (Baltimore: Johns Hopkins University Press, 2006), 59.

34. Amir Efrati, "The Home Front: McMansion Expansion," *Wall Street Journal*, August 12, 2005, W8.

35. Robert E. Lang and Karen A. Danielsen, "Monster Homes," *Planning* (May 2002): 2–7.

36. Haya El Nasser, "Teardowns' Have Critics Torn Up," *USA Today*, June 28, 2006, 3A.

37. Efrati, "Home Front."

38. Therese Howe, "Ritz's Creighton Farms Opens for Home Sales," *Leesburg Today*, October 20, 2006, 23.

39. Adrienne Schmitz and R. Bird Anderson Jr., *Residential Development Handbook* (Washington, D.C.: Urban Land Institute, 2003).

40. Urban Land Institute, *Housing for Emerging Niche Markets*, InfoPacket No. 3007 (Washington, D.C.: Urban Land Institute, 2006).

41. Mark D. Bjaelland, Michelle Maley, et al., "The Quest for Authentic Place: The Production of Suburban Alternatives in Minnesota's St. Croix Valley," *Urban Geography* 27 (2006): 253–270.

42. Urban Land Institute, *Great Planned Communities* (Washington, D.C.: ULI, 2002); see also Lang and Le Furgy, *Boomburbs*, chapter 5.

43. Aliante, "Aliante: An Exciting New Community" (company sales literature), 2003.

44. Lang and Le Furgy, *Boomburbs*, 107

45. Brian Skoloff, "Founder Planning a Strict New City," *The Ledger*, March 1, 2006, http://www.theledger.com/apps/pbcs.dll/article?AID=/20060301/NEWS/603010366&SearchID=73269870269582 (accessed June 16, 2007). The development's web site is *http://www.avemaria.com/*

CHAPTER 5   *Comfortably Numb: Degenerate Utopias and Their Evangelistic Consultants*

1. Edward Relph, *The Modern Urban Landscape* (London: Croom Helm, 1987), 130.

2. David Harvey, *Spaces of Hope* (Berkeley: University of California Press, 2000); Louis Marin, *Utopics: The Semiological Play of Textual Spaces* (Amherst, N.Y.: Prometheus Books, 1990).

3. Joe Cullen and Paul Knox, "'The Triumph of the Eunuch': Planners, Urban Managers, and the Suppression of Political Opposition," *Urban Affairs Quarterly* 17 (1981): 149–172.

4. Paul Knox, "Town Planning and the Internal Survival Mechanisms of Urbanized Capitalism," *Area* 13 (1981): 183–188.

5. Paul Knox and Joe Cullen, "Planners as Urban Managers," *Environment & Planning A* 13 (1981): 885–898.

6. The "can-do" heroes include Robert Moses in New York; Edward J. Logue, who worked for Richard Lee's administration in New Haven; Edmund N. Bacon in Philadelphia; Dave Loeks in Minneapolis–St. Paul; and William Ryan Drew in Milwaukee.

7. Eran Ben-Joseph, *The Code of the City: Standards and the Hidden Language of Place Making* (Cambridge, Mass.: MIT Press, 2005).

8. Nan Ellin, *Integral Urbanism* (New York: Routledge, 2006).

9. Kim Dovey, *Framing Places: Mediating Power in Built Form* (New York: Routledge, 1999).

10. Bernard Tschumi, *Architecture and Disjunction* (Cambridge, Mass.: MIT Press, 1994), 174.

11. Rem Koolhaas, Stefano Boeri, et al., *Mutations: Harvard Project on the City* (Barcelona : ACTAR; Bordeaux, France: Arc en rêve centre d'architecture, 2001).

12. For example, the *Journal of Urban Design* 1 (1994) and *Opolis: An International Journal of Suburban Studies* 1 (2004).

13. Mary Peck, in her presidential speech to the 2005 meetings of the American Planning Association in San Francisco, proclaimed a "golden age of planning" where "people are building what we espouse," adding, "People are starting to use our language"— a language that emphasizes values such as "authenticity" and "sense of place." Philip Langdon, reporting on the speech in the April/May 2005 issue of *New Urban News* (http://www.newurbannews.com/AprMay05.html [accessed June 18, 2007]), peevishly suggested that planners were jumping on a new urbanist bandwagon.

14. This section draws on Paul Knox and Linda McCarthy, *Urbanization*, 2nd ed. (Upper Saddle River, N.J.: Prentice Hall, 2005), 143–145.

15. Ann Forsyth, *Reforming Suburbia: The Planned Communities of Irvine, Columbia, and the Woodlands* (Berkeley: University of California Press, 2005), 2.
16. Ibid.
17. R. J. Burby and S. Weiss, *New Communities USA* (Lexington Mass.: Lexington Books, 1976).
18. This section draws on Nan Ellin, *Postmodern Urbanism* (Oxford: Blackwell, 1996).
19. See Paul Goodman, *Communitas: Means of Livelihood and Ways of Life* (New York: Random House, 1960); Jane Jacobs, *The Death and Life of Great American Cities: The Failure of Town Planning* (New York: Vintage, 1961); Richard Sennett, *The Uses of Disorder: Personal Identity and City Life* (New York: Random House, 1970); Richard Sennett, *The Conscience of the Eye: The Design and Social Life of Cities* (New York: Knopf, 1990).
20. Kevin Lynch, *Image of the City* (Cambridge, Mass.: MIT Press, 1960); Christopher Alexander, *A Pattern Language: Towns, Buildings, Construction* (New York: Oxford University Press, 1977); Christopher Alexander, *The Timeless Way of Building* (New York: Oxford University Press, 1979).
21. Robert Delevoy, quoted in Ellin, *Postmodern Urbanism*, 10.
22. Ferdinand Tönnies argued that two basic forms of human association could be recognized in all cultural systems (*Community and Society*, trans. C. P. Loomis [1887; reprint, New York: Harper, 1963]). The first of these, *Gemeinschaft*, he related to an earlier period in which the basic unit of organization was the family or kin-group, with social relationships characterized by depth, continuity, cohesion, and fulfillment. The second, *Gesellschaft*, was seen as the product of urbanization and industrialization that resulted in social and economic relationships based on rationality, efficiency, and contractual obligations among individuals whose roles had become specialized. See chapter 7.
23. Bernard Rudofsky, *Architecture Without Architects: An Introduction to Non-Pedigreed Architecture* (New York: Museum of Modern Art, 1964).
24. Jacques Ribaud, quoted in Ellin, *Postmodern Urbanism*, 31.
25. Colin Rowe, "Collage City," *Architectural Review* (August 1975): 65–91.
26. Robert Venturi, Denise Scott Brown, and Steven Izenour, *Learning from Las Vegas* (Cambridge, Mass.: MIT Press, 1972); Robert Venturi, *Complexity and Contradiction in Architecture* (New York: Museum of Modern Art, 1966).
27. Guy Debord, *The Society of the Spectacle* (New York: Zone Books, 1993); Fredric Jameson, *The Cultural Turn: Selected Writings on the Postmodern, 1983–1998* (London: Verso, 1998); Fredric Jameson, *Postmodernism; or, The Cultural Logic of Late Capitalism* (Durham, N.C.: Duke University Press, 1991).
28. Edward Blakely and Mary Gail Snyder, *Fortress America: Gated Communities in the United States* (Washington, D.C.: Brookings Institution Press, 1999).
29. Kevin Romig, "Upper Sonoran Lifestyles: Gated Communities in Scottsdale, Arizona," *City and Community* 4 (2005): 67–86; Georjeanna Wilson-Dogenges, "An Exploration of Sense of Community and Fear of Crime in Gated Communities," *Environment and Behavior* 32 (2000): 597–611.
30. Andrew Kirby, Sharon L. Harlan, et al., "Examining the Significance of Housing Enclaves in the Metropolitan United States of America," *Housing, Theory, and Society* 23 (2006): 30. See also Thomas W. Sanchez, Robert E. Lang, and Dawn M. Dhavale, "Security versus Status? A First Look at the Census's Gated Community Data," *Journal of Planning Education and Research* 24 (2005): 281–291.

31. Setha Low, *Behind the Gates: Life, Security, and the Pursuit of Happiness in Fortress America* (New York: Routledge, 2003).
32. Sanchez et al., "Security versus Status," 283.
33. U.S. Department of Housing and Urban Development Office of Policy Development and Research, *American Housing Survey for the United States 2005* (Washington, D.C.: U.S. Department of Housing and U.S. Department of Commerce, 2006), Table 2–8.
34. Blakely and Snyder, *Fortress America*, 18.
35. Quoted in Tschumi, *Architecture and Disjunction*, 227.
36. Ellin, *Postmodern Urbanism*, 75.
37. The Ahwahnee Principles are available on California's Local Government Commission web site: *http://www.lgc.org/ahwahnee/principles.html* (accessed June 18, 2007).
38. Linda Hales, "Little House on the Drawing Board," *Washington Post*, October 18, 2006, C1.
39. The quote is attributed to Ken Gancarczyk, an executive of KB Home, in ibid.
40. Quoted in Ellin, *Postmodern Urbanism*, 85.
41. Available at *http://www.privatecommunities.com/visit/index.htm?community_id= 977&community_name=Withers%20Preserve&preview=0&link_location=top* (accessed January 12, 2007).
42. Available at *http://www.reservecypresshills.com/* (accessed January 12, 2007).
43. United Nations Centre for Human Settlements, *Global Report on Human Settlements, 2001* (London: Earthscan, 2001), 4.
44. Paul Knox, "Capital, Material Culture, and Socio-Spatial Differentiation," in *The Restless Urban Landscape*, ed. P. L. Knox (Englewood Cliffs, N.J.: Prentice Hall, 1993), 1–32.
45. Debord, *The Society of the Spectacle*. Debord's concept of the spectacle is integrally connected to the concept of separation and passivity, for in submissively consuming spectacles, one is estranged from actively producing one's life.
46. Douglas Kellner, "Media Culture and the Triumph of the Spectacle," *Fast Capitalism* 1.1 (2005).
47. A. Forty and H. Moss, "A Housing Style for Troubled Consumers," *Architectural Review* 167 (1980): 22–28.
48. Quoted in Ellin, *Postmodern Urbanism*, 74.
49. Peter Calthorpe, "New Urbanism: Principles or Style?" in *New Urbanism. Peter Calthorpe vs. Lars Lerup*, Michigan Debates on Urbanism, vol. 2 (Ann Arbor, Mich.: Taubman College of Architecture + Planning, 2005), 17.
50. Umberto Eco describes the hyperreal as that which is absolutely fake in order to be better than the real. See Umberto Eco, *Travels in Hyperreality* (New York: Harcourt Brace Jovanovich, 1986).
51. Ellin, *Integral Urbanism*, 99.
52. Martin Heidegger, *Poetry, Language, Thought* (New York: Harper & Row, 1971).
53. Jean Baudrillard, *Simulacra and Simulation*, trans. Sheila F. Glaser (Ann Arbor: University of Michigan Press, 1994).
54. Janet Abu-Lughod, "Disappearing Dichotomies," *Traditional Dwellings and Settlements Review* 3 (1992). See also D. Upton, "The Tradition of Change," *Traditional Dwellings and Settlements Review* 5 (1993); Nezar AlSayyad, ed., *The End of Tradition?* (New York: Routledge, 2004).

55. Richard Sennett, "The Search for a Place in the World," in *Architecture of Fear*, ed. Nan Ellen (New York: Princeton Architectural Press, 1997), 61–72.

56. Peter Katz, ed., *The New Urbanism: Toward an Architecture of Community* (New York: McGraw-Hill, 1994); James H. Kunstler, *The Geography of Nowhere: The Rise and Decline of America's Man-Made Landscape* (New York: Simon & Schuster, 1993); James H. Kunstler, *Home from Nowhere: Remaking Our Everyday World for the Twenty-first Century* (New York: Simon & Schuster, 1996).

57. See http://www.cnu.org/charter.

58. Douglas Kelbaugh, "Three Paradigms: New Urbanism, Everyday Urbanism, Post Urbanism," Periferia Architecture and Design in the Caribbean, http://www.periferia.org/3000/3paradigms.html (accessed August 1, 2007).

59. Jill Grant, *Planning the Good Community: New Urbanism in Theory and Practice* (New York: Routledge, 2006), 11.

60. Though at the time of writing new urbanism does not feature among the "Knowledge Communities" of the American Institute of Architects.

61. David Mohney and Keller Easterling, eds., *Seaside: Making a Town in America* (Princeton, N.J.: Princeton Architectural Press, 1991).

62. Andrew Ross, *The Celebration Chronicles: Life, Liberty, and the Pursuit of Property Value in Disney's New Town* (New York: Ballantine Books, 1999); Douglas Frantz and Catherine Collins, *Celebration, U.S.A.: Living in Disney's Brave New Town* (New York: Henry Holt, 1999).

63. Ross, *Celebration Chronicles*, 19–20.

64. J. Kim and R. Kaplan, "Physical and Psychological Factors in Sense of Community: New Urbanist Kentlands and Nearby Orchard Village," *Environment and Behavior* 36 (2004); 313–340; Grant, *Planning the Good Community*.

65. Jerry W. Jackson " 'It's a Good Thing,' Home Builder Thinks," *Orlando Sentinel*, January 5, 2007, http://www.orlandosentinel.com/news/local/orange/orl-marthahomes0507jan05,0,3499019.story?coll=orl-news-headlines-orange (accessed January 8, 2007).

66. Alex Marshall, *How Cities Work. Suburbs, Sprawl, and the Roads Not Taken* (Austin: University of Texas Press, 2003), 12.

67. Edward Robbins, "New Urbanism," in *Shaping the City. Studies in History, Theory, and Urban Design*, ed. Edward Robbins and Rodolphe El-Khoury (New York: Routledge, 2004), 228.

68. J. Kim, "Perceiving and Valuing Sense of Community in a New Urbanist Development: The Case of Kentlands," *Journal of Urban Design* 12 (2007): 203–230.

69. David Harvey, "The New Urbanism and the Communitarian Trap: On Social Problems and the False Hope of Design," *Harvard Design Magazine* (Winter/Spring 1997): 68–69; Paul W. Clarke, "The Ideal of Community and Its Counterfeit Construction," *Journal of Architectural Education* 58 (2005): 44.

70. For the most comprehensive treatments of new urbanism and extended evaluations of the critique of new urbanism see Grant, *Planning the Good Community*, and Emily Talen, *New Urbanism and American Planning* (New York: Routledge, 2005).

71. Alex Krieger, "Whose Urbanism?" *Architecture* (November 1998): 75.

72. Peter Calthorpe, *The Next American Metropolis* (New York: Princeton Architectural Press, 1993), 10.

73. Grant, *Planning the Good Community*, 51.

74. Talen, *New Urbanism and American Planning*.

CHAPTER 6   *The Politics of Privatism*

1. Evan McKenzie, *Privatopia: Homeowner Associations and the Rise of Residential Private Government* (New Haven: Yale University Press, 1994), 177.
2. Torin Monahan, "Electronic Fortification in Phoenix: Surveillance Technologies and Social Regulation in Residential Communities," *Urban Affairs Review* 42 (2006): 169–192. This article discusses the role of surveillance in the emerging neoliberal society.
3. Jamie Peck and Adam Tickell, "Neoliberalizing Space," in *Spaces of Neoliberaism: Urban Restructuring in North America and Western Europe*, ed. Neil Brenner and Nik Theodore (Oxford: Blackwell, 2002), 33–57; Jason Hackworth, *The Neoliberal City: Governance, Ideology, and Development in American Urbanism* (Ithaca, N.Y.: Cornell University Press, 2006); Helga Leitner, Jamie Peck, and Eric Sheppard, eds., *Contesting Neoliberalism: Urban Frontiers* (New York: Guilford Press, 2007); Margaret Kohn, *Brave New Neighborhoods: The Privatization of Public Space* (New York: Routledge, 2004).
4. Oren Dorell, "Some New Cities Outsource City Hall," *USA Today*, September 15, 2006, A1; Geoffrey Segal, "Georgia City Shows Florida How To Cut Costs," *Tampa Tribune*, June 12, 2007, http://www.tbo.com/news/opinion/commentary/MGBGHS-DLT2F.html (accessed June 26, 2007).
5. Neil Smith, *The New Urban Frontier* (New York: Routledge, 1996). "Revanchism" derives from the French term for revenge, used since the 1870s in relation to the impulse to reverse territorial losses by a country as the result of a war.
6. Neil Brenner and Nik Theodore, "Cities and the Geography of 'Actually Existing Neoliberalism,'" in *Spaces of Neoliberalism: Urban Restructuring in North America and Western Europe*, ed. Neil Brenner and Nik Theodore (Oxford: Blackwell, 2002), 21.
7. For the origins of this interpretation, see Edward Banfield and J. Q. Wilson, *City Politics* (Cambridge, Mass.: Harvard University Press, 1963).
8. Jürgen Habermas, *The Structural Transformation of the Public Sphere* (Cambridge, Mass.: MIT Press, 1989).
9. Edward Blakely and Mary Gail Snyder, *Fortress America: Gated Communities in the United States* (Washington, D.C.: Brookings Institution Press, 1999), 139–140.
10. Setha Low, *Behind the Gates: Life, Security, and the Pursuit of Happiness in Fortress America* (New York: Routledge, 2003).
11. Stephen E. Barton and Carol J. Silverman, eds., *Common Interest Communities: Private Governments and the Public Interest* (Berkeley, Calif.: Institute of Governmental Studies Press, University of California, 1994).
12. Mike Davis, *City of Quartz* (London: Verso, 1990), 169–170.
13. McKenzie, *Privatopia*, 10.
14. Ibid., 12.
15. Adrienne Schmitz, R. Bird Anderson, et al., *Residential Development Handbook* (Washington, D.C.: Urban Land Institute, 2004), 185.
16. Ibid., 159.
17. See, for example, Andrew Jonas and David Wilson, eds., *The Urban Growth Machine: Critical Perspectives Two Decades Later* (Albany: State University of New York Press, 1999); J. R. Feagin, *The New Urban Paradigm: Critical Perspectives on the City* (Lanham, Md.: Rowman and Littlefield, 1997); M. Lauria, *Reconstructing*

Urban Regime Theory (Beverly Hills, Calif.: Sage, 1996); C. Rutheiser, *Imagineering Atlanta* (New York: Verso, 1996).

18. Robert H. Nelson, *Private Neighborhoods and the Transformation of Local Government* (Washington, D.C.: Urban Institute Press, 2005), 195 and xviii.

19. Evan McKenzie, "Constructing the *Pomerium* in Las Vegas: A Case Study of Emerging Trends in American Gated Communities," in *Gated Communities*, ed. Rowland Atkinson and Sarah Blandy (New York: Routledge, 2006), 1; see also William A. Fischel, "The Rise of Private Neighborhood Associations: Revolutions or Evolutions," in *The Property Tax, Land Use, and Land Use Regulation*, ed. D. Netzer (Northampton, Mass.: Edgar Elgar, 2003).

20. Nelson, *Private Neighborhoods*, 3.

21. Alex Marshall, *How Cities Work: Suburbs, Sprawl, and the Roads Not Taken* (Austin: University of Texas Press, 2000); Blakely and Snyder, *Fortress America*; McKenzie, "Constructing the *Pomerium*," 3.

22. McKenzie, *Privatopia*, 16.

23. Ibid., 15.

24. Timothy B. Wheeler, "Abiding by Rules of Your Neighbor," *Baltimore Sun*, December 25, 2005, *http://www.baltimoresun.com/news/local/balte.md. association25dec25,1,743023.story?ctrack=1&cset=true* (accessed June 18, 2007).

25. Ibid.

26. Lakiesha McGhee, "Home Groups' Rules Assailed State Oversight and Fees Are Sought to Rein in Homeowner Associations' Power," *Sacramento Bee*, January 25, 2005, B1.

27. Laura Manseras, "Chalk One Up for Homeowners," *New York Times*, March 12, 2006, Section 14NJ, 1.

28. Rob Lang and Jennifer LeFurgy, *Boomburbs* (Washington, D.C.: Brookings Institution Press, 2007).

29. Emmet Pierce, "Associations Are Too Quick on the Trigger, Critics Claim," *San Diego Union-Tribune*, April 10, 2005, 1–6.

30. McKenzie, *Privatopia*, 21.

31. See Nelson, *Private Neighborhoods*, for detailed argumentation of this perspective.

32. Lang and LeFurgy, *Boomburbs*, chapter 6.

33. See, for example, A. Dan Tarlock, "Toward a Revised Theory of Zoning," in *Land Use Controls Annual*, ed. F. S. Bangs (Chicago: American Society of Planning Officials, 1972), 141–150; Robert C. Ellickson, "Suburban Growth Controls: An Economic and Legal Analysis," *Yale Law Journal* 86 (January 1977): 385-511; Bernard H. Siegan, *Land Use without Zoning* (Lexington, Mass.: Lexington Books, 1972).

34. William A. Fischel, *The Homevoter Hypothesis: How Home Values Influence Local Government Taxation, School Finance, and Land-Use Policies* (Cambridge, Mass.: Harvard University Press, 2001).

35. Ken Newton, "Conflict Avoidance and Conflict Suppression," in *Urbanization and Conflict in Market Societies*, ed. Kevin Cox (London: Methuen, 1978), 84.

36. Tom Daniels, *When City and Country Collide: Managing Growth in the Metropolitan Fringe* (Washington, D.C.: Island Press, 1999), xiv.

37. Dolores Hayden, *Building Suburbia: Green Fields and Urban Growth, 1820–2000* (New York: Pantheon, 2003).

38. Jon Gertner, "Chasing Ground," *New York Times Magazine*, October 16, 2005, http://www.nytimes.com/2005/10/16/magazine/16brothers.html?pagewanted=1 (accessed January 25, 2007).

39. James Duncan and Nancy Duncan, *Landscapes of Privilege: The Politics of the Aesthetic in an American Suburb* (New York: Routledge, 2004), 85.

40. Jonathan Levine, *Zoned Out. Regulation, Markets, and Choices in Transportation and Metropolitan Land-Use* (Washington, D.C.: Resources for the Future, 2006); Edward L. Glaeser and Joseph Gyourko, *The Impact of Zoning on Housing Affordability*, Working Paper 8835 (Cambridge, Mass.: National Bureau of Economic Research, 2002).

41. Arthur C. Nelson, "How Do We Know Smart Growth When We See It?" in *Smart Growth. Form and Consequences*, ed. Terry Szold and Armando Carbonell (Cambridge, Mass.: Lincoln Institute of Land Policy, 2002), 82–101.

42. *http://www.smartgrowthamerica.org/* (Smart Growth of America, accessed June 18, 2007).

43. *http://www.smartgrowth.org/sgn/default.asp* (Smart Growth Online, accessed June 18, 2007).

44. F. Kaid Benfield, Jutka Terris, and Nancy Vorsanger, *Solving Sprawl: Models of Smart Growth in Communities across America* (Washington, D.C.: Island Press, 2001), 4.

45. See, for example, Wendell Cox and Ronald Utt, "Housing Affordability: Smart Growth Abuses Are Creating a 'Rent Belt' of High-Cost Areas," Heritage Foundation, *Backgrounder #1999*, http://www.heritage.org/Research/SmartGrowth/bg1999.cfm (accessed June 18, 2007).

46. Peter Whoriskey, "Planners' Brains vs. Public's Brawn," *Washington Post*, August 10, 2004, A01.

47. The remainder of this section is based closely on these articles. See, for example, Michael Laris and David S. Fallis, "Influence of Developers, Allies Runs Deep," *Washington Post*, January 21, 2007, A01, http://www.washingtonpost.com/wp-dyn/content/article/2007/01/20/AR2007012001493.html?sub=AR (accessed June 18, 2007).

CHAPTER 7   *Material Culture and Society in Metroburbia*

1. David Brooks, *On Paradise Drive* (New York: Simon & Schuster, 2004), 248.

2. Ibid., 247.

3. Colin Campbell, *The Romantic Ethic and the Spirit of Modern Consumerism* (Oxford: Blackwell, 1987).

4. James Duesenberry, *Income, Saving, and the Theory of Consumer Behavior* (Cambridge, Mass.: Harvard University Press, 1949); see also John Kenneth Galbraith, *The Affluent Society* (Boston: Houghton Mifflin, 1976); Lizabeth Cohen, *A Consumers' Republic: The Politics of Mass Consumption in Postwar America* (New York: Vintage, 2003).

5. Guy Debord, *La Société du Spectacle* (Paris: Buchet-Chastel, 1967).

6. Jean Baudrillard, *The System of Objects* (Bath, Eng.: Bath Press, 1968).

7. Galbraith, *The Affluent Society*; Agnes Heller, "Existentialism, Alienation, Postmodernism: Cultural Movements as Vehicles of Change in Patterns of Everyday

Life," in *Postmodern Conditions*, ed. A. Milner, P. Thompson, and C. Worth (New York: Berg, 1990), 1–13.

8. David Ley, "Modernism, Postmodernism, and the Struggle for Place," quoted in Paul Knox, "Capital, Material Culture, and Socio-Spatial Differentiation," in *The Restless Urban Landscape*, ed. Paul L. Knox (Englewood Cliffs, N.J.: Prentice-Hall 1993), 23.

9. K. Butler, "Pate Poverty: Downwardly Mobile Baby Boomers Lust after Luxury," *Utne Reader* (September/October 1989), 77.

10. Barbara Ehrenreich, *Fear of Falling* (New York: Pantheon, 1989), 229.

11. Juliet Schor, *The Overspent American* (New York: Harper Perennial, 1998), 9.

12. Robert H. Frank, *The Winner-Take-All Society: Why the Few at the Top Get So Much More Than the Rest of Us* (New York: Penguin, 1996).

13. Pierre Bourdieu, *Distinction: A Social Critique of the Judgement of Taste* (London: Routledge & Kegan Paul, 1984).

14. Scott Lash and John Urry, *Economies of Signs and Spaces: After Organized Capitalism* (Cambridge: Cambridge Policy Press, 1992); Ulrich Beck, Anthony Giddens, and Scott Lash, *Reflexive Modernization: Politics, Tradition, and Aesthetics in the Modern Social Order* (Stanford, Calif.: Stanford University Press, 1994).

15. Edward Soja, *Postmetropolis* (Oxford: Blackwell, 2000), 276.

16. U.S. Census Bureau, *Current Population Survey, Annual Social and Economic Supplements, http://www.census.gov/hhes/www/income/histinc/inchhtoc.html* (accessed June 20, 2007).

17. Robert H. Frank, *Luxury Fever* (Princeton, N.J.: Princeton University Press, 2000); John De Graaf, David Wann, and Thomas H. Naylor, *Affluenza: The All-Consuming Epidemic* (San Francisco: Berrett Koehler, 2001).

18. Robert H. Frank, *Richistan: A Journey through the American Wealth Boom and the Lives of the New Rich* (New York: Crown, 2007).

19. Geographic consumer profiles based on consumer surveys linked to statistical analyses of socioeconomic data by zip code.

20. Details of the VALS™ survey are available at http://www.sric-bi.com/VALS/presurvey.shtml (accessed June 20, 2007).

21. Paul Knox, "Vulgaria: The Re-Enchantment of Suburbia," *Opolis* 1 (2005): 34–47.

22. SRI Consulting Business Intelligence, *Understanding U.S. Consumers* (Menlo Park, Calif.: SRI-BI, 2003).

23. Virginia Postrel, *The Substance of Style* (New York: HarperCollins, 2003), 103.

24. M. Featherstone, *Consumer, Culture and Post-modernism* (London: Sage Publications, 1991); Colin Campbell, "The Sociology of Consumption," in *Acknowledging Consumption*, ed. Daniel Miller (New York: Routledge, 1995), 96–126; Mark Paterson, *Consumption and Everyday Life* (New York: Routledge, 2006).

25. Agnes Heller, "Existentialism, Alienation, Postmodernism: Cultural Movements as Vehicles of Change in Patterns of Everyday Life," in *Postmodern Conditions*, ed. A. Milner, P. Thompson, and C. Worth (New York: Berg, 1990), 1–13.

26. Guy Debord, *Comments on the Society of the Spectacle* (London: Verso, 1990), para. 167.

27. Kirsten Gram-Hanssen and Claus Bech-Danielsen, "House, Home and Identity from a Consumption Perspective," *Housing, Theory and Society* 21 (2004): 17–26.

28. John Archer, *Architecture and Suburbia: From English Villa to American Dream House, 1690–2000* (Minneapolis: University of Minnesota Press, 2005), 366.

29. G. McCracken, *Culture and Consumption: New Approaches to the Symbolic Character of Consumer Goods and Activities* (Bloomington: Indiana University Press, 1988), 124.

30. Robert Sack, "The Consumer's World: Place as Context," *Annals, Association of American Geographers* 78 (1988): 642–665.

31. L. Shames, "What a Long, Strange (Shopping) Trip It's Been: Looking Back at the 1980s," *Utne Reader* (September/October 1989): 66.

32. John Eyles, "Housing Advertisements as Signs: Locality Creation and Meaning Systems," *Geografiska Annaler* 69B (1987): 95.

33. Michel Maffesoli, *The Time of the Tribes. The Decline of Individualism in Mass Society* (London: Sage, 1996); see also Rob Shields, "The Individual, Consumption Cultures, and the Fate of Community," in *Lifestyle Shopping: The Subject of Consumption*, ed. R. Shields (New York: Routledge, 1992), 99–113.

34. Bourdieu, *Distinction*, 173. See also Pierre Bourdieu, "Habitus," in *Habitus: A Sense of Place*, ed. Jean Hillier and Emma Rooksby (Aldershot, Eng.: Ashgate, 2002), 27–34.

35. Pierre Bourdieu, *In Other Words* (Cambridge: Polity, 1990), 113.

36. Thorstein Veblen, *The Theory of the Leisure Class* (New York: Macmillan, 1899).

37. See, for example, Paul L. Knox and Steven Pinch, *Urban Social Geography*, 5th ed. (London: Pearson, 2006).

38. See *http://www.claritas.com/claritas/Default.jsp?ci=3&si=4&pn=prizmne* (accessed June 20, 2007).

39. Richard Sennett, *The Conscience of the Eye: The Design and Social Life of Cities* (New York: Knopf, 1990), xii.

40. Ferdinand Tönnies, *Community and Society*, trans. C. P. Loomis (1887; reprint, New York: Harper, 1963).

41. Graham Day, *Community and Everyday Life* (New York: Routledge, 2006); Knox and Pinch, *Urban Social Geography*, chapter 9.

42. Herbert Gans, *The Urban Villagers* (New York: Free Press, 1962).

43. Lewis Mumford, *The Culture of Cities* (London: Secker and Warburg, 1940), 215.

44. See, for example, R. S. Lynd and H. M. Lynd, *Middletown* (New York: Harcourt Brace Javanovich, 1956), and W. L. Warner and P. S. Lunt, *The Social Life of a Modern Community* (New Haven: Yale University Press, 1941). Further sociological work such as William Whyte's *The Organization Man* (New York: Doubleday, 1956) and M. Stein's *The Eclipse of Community* (New York: Harper & Row, 1960) reinforced the image of the suburbs as an area of loose-knit, secondary ties where lifestyles were focused squarely on the nuclear family's pursuit of money, status, and consumer durables and the privacy in which to enjoy them.

45. Herbert Gans, *The Levittowners: Ways of Life and Politics in a New Suburban Community* (New York: Vintage, 1967).

46. Miller McPherson, Lyn Smith-Lovin, and Matthew Brashears, "Social Isolation in America: Changes in Core Discussion Networks over Two Decades," *American Sociological Review* 71 (2006): 353–375.

47. Lisa McGurr, *Suburban Warriors: The Origins of the New American Right* (Princeton, N.J.: Princeton University Press, 2001).

48. Edward Blakely and Mary Gail Snyder, *Fortress America: Gated Communities in the United States* (Washington, D.C.: Brookings Institution Press, 1999), 130.

49. Robert Putnam, *Bowling Alone: The Collapse and Revival of American Community* (New York: Simon & Schuster, 2000).

50. Emily Talen, "Sense of Community and Neighborhood Form: An Assessment of the Social Doctrine of New Urbanism," *Urban Studies* 36 (1999): 1361.

51. See, for example, B. Brown and V. Cropper, "New Urban and Standard Suburban Subdivisions: Evaluating Psychological and Social Goals," *Journal of the American Planning Association* 67 (2001): 402–419; H. Lund, "Testing the Claims of New Urbanism: Local Access, Pedestrian Travel, and Neighboring Behaviors," *Journal of the American Planning Association* 69 (2003): 414–29; J. Nasar, "Does Neotraditional Development Build Community?" *Journal of Planning Education and Research* 23 (2003): 58–68; T. Brindley, "The Social Dimension of the Urban Village: A Comparison of Models for Sustainable Urban Development," *Urban Design International* 8 (2003): 53–65.

52. Jill Grant, *Planning the Good Community: New Urbanism in Theory and Practice* (New York: Routledge, 2006).

53. Ibid., 211.

CHAPTER 8   *Vulgaria: Moral Landscapes at the Leading Edges of the New Metropolis*

1. Sharon Zukin, *Landscapes of Power: From Detroit to Disney World* (Berkeley: University of California Press, 1991).

2. W.J.T. Mitchell, ed., *Landscapes and Power* (Chicago: University of Chicago Press, 1994); James and Nancy Duncan, *Landscapes of Privilege: The Politics of the Aesthetic in an American Suburb* (New York: Routledge, 2004); Zukin, *Landscapes of Power*.

3. Don Mitchell, *Cultural Geography: A Critical Introduction* (Oxford: Blackwell, 2000).

4. Kim Dovey, *Framing Places: Mediating Power in Built Form* (New York: Routledge, 2000), 2, 45.

5. Don Mitchell, "Landscape and Surplus Value: The Making of the Ordinary in Brentwood, California," *Environment & Planning D: Society and Space* 12 (1994): 10.

6. Nigel Thrift, "Transurbanism," *Urban Geography* 25 (2004): 726.

7. Colin Campbell, *The Romantic Ethic and the Spirit of Modern Consumerism* (Oxford: Blackwell, 1989); David Brooks, *On Paradise Drive* (New York: Simon & Schuster, 2004).

8. Paul L. Knox, "Schlock and Awe," *American Interest* 2 (2007): 58–67.

9. Guy Debord, *The Society of the Spectacle* (New York: Zone Books, 1994); David Harvey, *Spaces of Global Capitalism: A Theory of Uneven Geographical Development* (New York: Verso, 2006).

10. Robert Venturi, Denise Scott Brown, and Steven Izenour, *Learning from Las Vegas: The Forgotten Symbolism of Architectural Form* (Cambridge, Mass.: MIT Press, 1977).

11. This section draws on Paul Knox and Kylie Johnson, "Learning from Las Vegas: Place Marketing and the Arts," *International Journal of the Arts in Society* 1 (2007). See also Michelle Ferrari and Stephen Ives, *Las Vegas. An Unconventional History* (New York: Bullfinch Press, 2005); Hal Rothman, *Neon Metropolis: How Las Vegas Started the Twenty-first Century* (London: Routledge, 2002); and James Twitchell, *Living It Up: America's Love Affair with Luxury* (New York: Simon & Schuster, 2002), 215–267.

12. Ferrari and Ives, *Las Vegas*, 32.

13. Ibid.

14. Simon Sykes, "Profile: Living Las Vegas," *Urban Land* 64 (2005).

15. William Fox, *In the Desert of Desire: Las Vegas and the Culture of Spectacle* (Reno: University of Nevada Press, 2005).

16. Peter Weibel, "Las Vegas, the City—A Place of Consumption in the Post-industrial Information Society," in *The Magic Hour: The Convergence of Art and Las Vegas*, ed. Alex Farquharson (Ostfildern, Ger.: Hatje Cantz, 2001).

17. Michael Sorkin, "Brand Aid; or, The Lexus and the Guggenheim (Further Tales of the Notorious B.I.G.ness)," *Harvard Design Magazine* 17 (2002): 2.

18. These definitions are from the Oxford Dictionary Online (*http://www.askoxford. com/dictionaries/?view=uk*), and the *Longman Dictionary of the English Language* (Harlow: Longman, 1984).

19. See chapter 7.

20. Robert Bocock, *Consumption* (London: Routledge, 1993); Lizabeth Cohen, *A Consumer's Republic: The Politics of Mass Consumption in Postwar America* (New York: Knopf, 2003); William Leach, *Land of Desire, Merchants of Power, and the Rise of a New American Culture* (New York: Vintage, 1993); Susan Strasser, *The Making of the American Mass Market* (New York: Pantheon, 1989); and Alan Latham, "American Dreams, American Empires, American Cities," *Urban Geography* 25 (2004): 788–791.

21. H. L. Mencken, *Prejudices: Sixth Series* (New York: Knopf, 1927).

22. Beppe Severgnini, *Ciao America: An Italian Discovers the U.S.* (New York: Broadway Books, 2001), 216.

23. Brooks, *On Paradise Drive*, 90.

24. Lisa Mahar, *American Signs: Form and Meaning on Route 66* (New York: Monacelli, 2002).

25. Juliet Schor, *The Overspent American: Upscaling, Downshifting, and the New Consumer* (New York: Basic Books, 1998), 12, 13.

26. Ibid., 80.

27. Brooks, *On Paradise Drive*, 5.

28. Ibid., 31.

29. Pierre Bourdieu, *Outline of a Theory of Practice* (Cambridge: Cambridge University Press, 1977), 188. Some social scientists have explored such "complicitous silence" in spatial patterns and relations. Yi-Fu Tuan, for example, has written about the moral readings of the relationship between the moral and the aesthetic in landscape in *Passing Strange and Wonderful: Aesthetics, Nature, and Culture* (Washington, D.C.: Island Press, 1993). David Ley has interrogated the meanings attached to cooperative housing: David Ley, "Co-Operative Housing As a Moral Landscape: Re-Examining 'the Postmodern City,'" in *Place/culture/representation*, ed. J. Duncan and D. Ley

(London: Routledge, 1993), 128–148. Most studies of the moral geography of urban landscapes have been historical in nature, with Victorian cities receiving most attention. See Mona Domosh, "The 'Women of New York': A Fashionable Moral Geography," *Environment and Planning D: Society and Space* 19 (2001): 573–592; Felix Driver, "Moral Geographies: Social Science and the Urban Environment in Mid-Nineteenth Century England," *Transactions, Institute of British Geographers* n.s. 13 (1988): 275–287; Peter Jackson, "Social Disorganization and Moral Order in the City," *Transactions, Institute of British Geographers* n.s. 9 (1984): 168–180; M. Ogborn and C. Philo, "Soldiers, Sailors, and Moral Locations in Nineteenth-Century Portsmouth," *Area* 26 (1994): 221–231; T. Ploszajska, "Moral Landscapes and Manipulated Spaces: Gender, Class, and Space in Victorian Reformatory Schools," *Journal of Historical Geography* 20 (1994): 413–429.

30. Knox, "Schlock and Awe."
31. Duncan and Duncan, *Landscapes of Privilege*, 70.
32. Dovey, *Framing Places*, 143.
33. Ibid., 145–146.
34. http://www.tollbrothers.com/DYOH.swf?plan=hampton_978, (accessed June 24, 2007).
35. *http://www.lilliputplayhomes.com/* (accessed June 23, 2007).
36. *http://www.lapetitemaison.com/ph.html* (accessed June 23, 2007).
37. Troy McMullen, "Mini-Me Mcmansion Sales on the Rise," *Oregonian*, January 2, 2007, http://www.oregonlive.com/business/oregonian/index.ssf?/base/business/1167549903146330.xml&coll=7 (accessed January 8, 2007).
38. George Ritzer, *Enchanting a Disenchanted World*, 2nd ed. (Thousand Oaks, Calif.: Pine Forge Press, 2005).
39. Dolores Hayden, *Building Suburbia: Green Fields and Urban Growth, 1820–2000* (New York: Pantheon, 2003), 176.
40. See William Kowinski, *The Malling of America* (New York: Morrow, 1985); Jon Goss, "The 'Magic of the Mall': An Analysis of Form, Function, and Meaning in the Contemporary Retail Built Environment," *Annals of the Association of American Geographers* 83 (1993):18–47; Jon Goss, "Once-upon-a-Time in the Commodity World: An Unofficial Guide to Mall of America," *Annals of the Association of American Geographers* 89 (1999): 45–75.
41. *http://www.mallatmillenia.com/experience/experience.htm* (accessed on June 13, 2007).
42. Ritzer, *Enchanting a Disenchanted World*, 143.
43. Suzanne Goldenberg, "Hollywood Finds Christ as Foxfaith Plans Series of Religious Movies," *Guardian*, September 20, 2006, *http://film.guardian.co.uk/news/story/0,,1876553,00.html* (accessed June 24, 2007).
44. James Twitchell, *Branded Nation: The Marketing of Megachurch, College, Inc., and Museumworld* (New York: Simon & Schuster, 2004), 56.
45. See http://hirr.hartsem.edu/index.html (accessed June 13, 2007). There are as many as 3,000 Catholic churches in the United States with congregations of 2,000 or more, but almost none of them really function like the Protestant megachurches described here. See also Jonathan Weiss and Randy Lowell, "Supersizing Religion: Megachurches, Sprawl, and Smart Growth," *St. Louis University Public Law Review* 21 (2002): 313–329; and Jonathan Mahler, "The Soul of the New Exurb," *New York Times*, March

27, 2005, http://www.nytimes.com/2005/03/27/magazine/327MEGACHURCH. html? pagewanted=1&ei=5070&en=7f15a67b54e70644&ex=1129694400&adxnnl=0 &oref=login&adxnnlx=1129550869–6ypWqhu6mLAE5ptAHXvgWQ   (accessed December 5, 2005).

46. Mahler, "Soul of the New Exurb."
47. Elizabeth Wilson, "The Rhetoric of Urban Space," *New Left Review* 209 (1995): 146–160.
48. Lewis Mumford, "Regions—To Live In," *Survey* 54 (1925): 151.

BIBLIOGRAPHY

Abbot, Carl. *The Metropolitan Frontier: Cities in the Modern American West.* Tucson: University of Arizona Press, 1993.

Abu-Lughod, Janet. "Disappearing Dichotomies." *Traditional Dwellings and Settlements Review* 3 (1992): 7–12.

Adams, James T. *The Epic of America.* Boston: Little, Brown, 1931.

Alexander, Christopher. *A Pattern Language: Towns, Buildings, Construction.* New York: Oxford University Press, 1977,

———. *The Timeless Way of Building.* New York: Oxford University Press, 1979.

AlSayyad, Nezar, ed. *The End of Tradition?* New York: Routledge, 2004.

Archer, John. *Architecture and Suburbia. From English Villa to American Dream House, 1690–2000.* Minneapolis: University of Minnesota Press, 2005.

Badger, Anthony J. *The New Deal: The Depression Years 1933–1940.* Chicago: Ivan R. Dee, 2002.

Ball, Michael. "The Built Environment and the Urban Question." *Environment & Planning D: Society and Space* 4 (1986): 447–464.

———. "The Organization of Property Development Professions." In *Development and Developers,* ed. Simon Guy and John Henneberry. Oxford: Blackwell, 2002.

Banfield, Edward. *The Unheavenly City: The Nature and the Future of Our Urban Crisis.* Boston: Little, Brown, 1968.

Banfield, Edward, and J. Q. Wilson. *City Politics.* Cambridge, Mass.: Harvard University Press, 1963.

Barthes, Roland. *Mythologies.* Trans. A. Lavers. London: Paladin, 1973.

Barton, Stephen E., and Carol J. Silverman, eds. *Common Interest Communities: Private Governments and the Public Interest.* Berkeley: Institute of Governmental Studies Press, University of California, 1994.

Baudrillard, Jean. *The Consumer Society: Myths and Structures.* London: Sage, 1998.

———. *For a Critique of the Political Economy of the Sign.* St. Louis: Telos, 1981.

———. *Simulacra and Simulation.* Trans. Sheila F. Glaser. Ann Arbor: University of Michigan Press, 1994.

———. *The System of Objects.* Bath, Eng.: Bath Press, 1968.

Beauregard, Robert. *When America Became Suburban.* Minneapolis: University of Minnesota Press, 2006.

Beck, Ulrich. *Risk Society: Towards a New Modernity.* London: Sage, 1992.

Beck, Ulrich, W. Bonss, and C. Lau. "The Theory of Reflexive Modernization." *Theory, Culture, and Society* 20 (2003): 1–33.

Beck, Ulrich, Anthony Giddens, and Scott Lash. *Reflexive Modernization: Politics, Tradition and Aesthetics in the Modern Social Order.* Stanford, Calif.: Stanford University Press, 1994.

Benfield, F. Kaid, Jutka Terris, and Nancy Vorsanger. *Solving Sprawl: Models of Smart Growth in Communities across America.* Washington, D.C.: Island Press, 2001.

Ben-Joseph, Eran. *The Code of the City: Standards and the Hidden Language of Place Making.* Cambridge, Mass.: MIT Press, 2005.

Berger, Bennett. *Working Class Suburb.* Berkeley: University of California Press, 1960.

Berube, Alan, Audrey Singer, Jill H. Wilson, and William H. Frey. *Finding Exurbia: America's Fast-Growing Communities at the Metropolitan Fringe.* Washington, D.C.: Brookings Institution Press, 2006.

Bjelland, Mark D., Lane Cowger, Michelle Maley, and Lisabeth Barajas. "The Quest for Authentic Place: The Production of Suburban Alternatives in Minnesota's St. Croix Valley." *Urban Geography* 27 (2006): 253–270.

Blakely, Edward, and Mary Gail Snyder. *Fortress America: Gated Communities in the United States.* Washington, D.C.: Brookings Institution Press, 1999.

Bocock, Robert. *Consumption.* London: Routledge, 1993.

Bondanella, Peter. *Umberto Eco and the Open Text.* Cambridge: Cambridge University Press, 1997.

Bourdieu, Pierre. *Distinction: A Social Critique of the Judgement of Taste.* London: Routledge & Kegan Paul, 1984.

———. *The Field of Cultural Production.* New York: Columbia University Press, 1993.

———. "Habitus." In *Habitus: A Sense of Place*, ed. Jean Hillier and Emma Rooksby. Aldershot, Eng.: Ashgate, 2002.

———. *In Other Words.* Cambridge: Polity, 1990.

———. *Outline of a Theory of Practice.* Cambridge: Cambridge University Press, 1977.

Brenner, Neil, and Nik Theodore. "Cities and the Geography of 'Actually Existing Neoliberalism.' " In *Spaces of Neoliberalism: Urban Restructuring in North America and Western Europe*, ed. Neil Brenner and Nik Theodore. Oxford: Blackwell, 2002.

Brindley, T. "The Social Dimension of the Urban Village: A Comparison of Models for Sustainable Urban Development." *Urban Design International* 8 (2003): 53–65.

Brooks, David. *On Paradise Drive.* New York: Simon & Schuster, 2004.

———. "Patio Man and the Sprawl People." *Weekly Standard* (12 August 2002): 19–21, 24–26, 28–29.

Brown, B., and V. Cropper. "New Urban and Standard Suburban Subdivisions: Evaluating Psychological and Social Goals." *Journal of the American Planning Association* 67 (2001): 402–419.

Bruegmann, Robert. *Sprawl: A Compact History.* Chicago: University of Chicago Press, 2005.

Bryant, C.G.A., and D. Jary, eds. *Giddens's Theory of Structuration: A Critical Appreciation.* London: Routledge, 1991.

Burby, R. J., and S. Weiss. *New Communities USA.* Lexington Mass.: Lexington Books, 1976.

Burchell, R., G. Lowenstein, W. R. Dolphin, C. C. Galley, A. Downs, S. Seskin, K. G. Still, and T. Moore. *Costs of Sprawl—2000.* Washington, D.C.: Transportation Research Board, National Research Council, 2002.

————. *Costs of Sprawl Revisited.* Washington, D.C.: National Academies Press, 1998.

Burchfield, Marcy, Henry G. Overman, Diego Puga, and Matthew A. Turner. "Causes of Sprawl: A Portrait from Space." *Quarterly Journal of Economics* (2006): 587–633.

Butler, K. "Paté Poverty: Downwardly Mobile Baby Boomers Lust after Luxury." *Utne Reader* (September/October 1989): 77–81.

Calthorpe, Peter. "New Urbanism: Principles or Style?" In *New Urbanism: Peter Calthorpe vs. Lars Lerup.* Michigan Debates on Urbanism, vol. 2. Ann Arbor, Mich.: Taubman College of Architecture + Planning, 2005.

————. *The Next American Metropolis.* New York: Princeton Architectural Press, 1993.

Campbell, Colin. *The Romantic Ethic and the Spirit of Modern Consumerism.* Oxford: Blackwell, 1989.

————. "The Sociology of Consumption." In *Acknowledging Consumption,* ed. Daniel Miller. New York: Routledge, 1995.

Center for Housing Policy. *A Heavy Load: The Combined Housing and Transportation Burdens of Working Families.* Washington, D.C.: Center for Housing Policy, 2006.

Chambers, John. *The Tyranny of Change: America in the Progressive Era, 1890–1920.* New Brunswick, N.J.: Rutgers University Press, 1992.

Checkoway, Barry. "Large Builders, Federal Housing Programs, and Postwar Suburbanization." *International Journal of Urban and Regional Research* 4 (1980): 21–45.

Clarke, Paul W. "The Ideal of Community and Its Counterfeit Construction." *Journal of Architectural Education* 58 (2005): 43–52.

Cohen, Lizabeth. *A Consumers' Republic: The Politics of Mass Consumption in Postwar America.* New York: Vintage, 2003.

Cox, Wendell. *War on the Dream: How Anti-Sprawl Policy Threatens the Quality of Life.* New York: Universe, 2006.

Cullen, Joe, and Paul L. Knox. " 'The Triumph of the Eunuch': Planners, Urban Managers, and the Suppression of Political Opposition." *Urban Affairs Quarterly* 17 (1981): 149–172.

Cutsinger, Jackie, and George Galster. "There Is No Sprawl Syndrome: A New Typology of Metropolitan Land Use Patterns." *Urban Geography* 27 (2006): 228–252.

Daniels, Tom. *When City and Country Collide: Managing Growth in the Metropolitan Fringe.* Washington, D.C.: Island Press, 1999.

Davis, Mike. *City of Quartz: Excavating the Future in Los Angeles.* London: Verso, 1990.

Day, Graham. *Community and Everyday Life.* New York: Routledge, 2006.

Dear, Michael. "Comparative Urbanism." *Urban Geography* 26 (2005): 247–251.

Dear, Michael, and Jennifer Wolch. "How Territory Shapes Social Life." In *The Power of Geography,* ed. M. Dear and J. Wolch. London: Unwin Hyman, 1989.

Debord, Guy. *Comments on the Society of the Spectacle.* London: Verso, 1990.

————. *The Society of the Spectacle.* New York: Zone Books, 1993.

de Certeau, Michel. *The Practice of Everyday Life.* Trans. S. Randall. Berkeley: University of California Press, 1984.

De Graaf, John, David Wann, and Thomas H. Naylor. *Affluenza: The All-Consuming Epidemic.* San Francisco: Berrett Koehler, 2001.

Domosh, Mona. "The 'Women of New York': A Fashionable Moral Geography." *Environment and Planning D: Society and Space* 19 (2001): 573–592.

Dovey, Kim. *Framing Places: Mediating Power in Built Form.* New York: Routledge, 1999.

Downing, Andrew Jackson. *Cottage Residences, Adapted to North America.* New York: Wiley and Putnam, 1844.

Driver, Felix. "Moral Geographies: Social Science and the Urban Environment in Mid-Nineteenth-Century England." *Transactions, Institute of British Geographers* n.s. 13 (1988): 275–287.

Duesenberry, James. *Income, Saving, and the Theory of Consumer Behavior.* Cambridge, Mass.: Harvard University Press, 1949.

Duncan, James S., and Nancy G. Duncan. *Landscapes of Privilege: The Politics of the Aesthetic in an American Suburb.* New York: Routledge, 2004.

Eco, Umberto. *The Open Work.* Cambridge, Mass.: Harvard University Press, 1989.

———. *Travels in Hyperreality.* New York: Harcourt Brace Jovanovich, 1986.

Edel, Matthew, Elliot D. Sclar, and Daniel Luria. *Shaky Palaces: Homeownership and Social Mobility in Boston's Suburbanization.* New York: Columbia University Press, 1984.

Ehrenreich, Barbara. *Fear of Falling.* New York: Pantheon, 1989.

Ellickson, Robert C. "Suburban Growth Controls: An Economic and Legal Analysis." *Yale Law Journal* 86 (1977): 385–511.

Ellin, Nan. *Integral Urbanism.* New York: Routledge, 2006.

Emerson, Ralph Waldo. *Nature.* Boston: James Munroe & Co., 1836.

The Exploding Metropolis. Garden City, N.Y.: Doubleday, 1958.

Eyles, John. "Housing Advertisements as Signs: Locality Creation and Meaning Systems." *Geografiska Annaler* 69B (1987): 93–105.

Feagin, J. R. *The New Urban Paradigm. Critical Perspectives on the City.* Lanham, Md.: Rowman and Littlefield, 1997.

Featherstone, M. *Consumer, Culture and Post-modernism.* London: Sage Publications, 1991.

Ferrari, Michelle, and Stephen Ives. *Las Vegas. An Unconventional History.* New York: Bullfinch Press, 2005.

Fischel, William A. *The Homevoter Hypothesis: How Home Values Influence Local Government Taxation, School Finance, and Land-Use Policies.* Cambridge, Mass.: Harvard University Press, 2001.

———. "The Rise of Private Neighborhood Associations: Revolutions or Evolutions." In *The Property Tax, Land Use, and Land Use Regulation*, ed. D. Netzer. Northampton, Mass.: Edward Elgar, 2003.

Fishman, Robert. "The American Planning Tradition." In *The American Planning Tradition*, ed. Robert Fishman. Washington, D.C.: Woodrow Wilson Center Press, 2000.

———. "Beyond Sprawl." In *Sprawl and Suburbia*, ed. William S. Saunders. Minneapolis: University of Minnesota Press, 2005.

———. *Bourgeois Utopias: The Rise and Fall of Suburbia.* New York: Basic Books, 1987.

———. *Urban Utopias in the Twentieth Century.* Cambridge, Mass.: MIT Press, 1987.

Flink, James J. *The Automobile Age.* Cambridge, Mass.: MIT Press, 1998.

Flint, Anthony. *This Land: The Battle over Sprawl and the Future of America.* Baltimore: Johns Hopkins University Press, 2006.

Fogelsong, Robert. *Bourgeois Nightmares: Suburbia 1870–1930.* New Haven: Yale University Press, 2005.

Forsyth, Ann. *Reforming Suburbia: The Planned Communities of Irvine, Columbia, and the Woodlands.* Berkeley: University of California Press, 2005.

Forty A., and H. Moss. "A Housing Style for Troubled Consumers." *Architectural Review* 167 (1980): 22–28.

Foucault, Michel. "Space, Power, and Knowledge." In *Rethinking Architecture*, ed. N. Leach. London: Routledge, 1997.

Fox, William. *In the Desert of Desire: Las Vegas and the Culture of Spectacle.* Reno: University of Nevada Press, 2005.

Frank, Robert H. *Luxury Fever.* Princeton, N.J.: Princeton University Press, 2000.

———. *Richistan: A Journey through the American Wealth Boom and the Lives of the New Rich.* New York: Crown, 2007.

———. *The Winner-Take-All Society: Why the Few at the Top Get So Much More Than the Rest of Us.* New York: Penguin, 1996.

Frantz, Douglas, and Catherine Collins. *Celebration, U.S.A.: Living in Disney's Brave New Town.* New York: Henry Holt, 1999.

Freund, David M. P. "Marketing the Free Market. State Intervention and the Politics of Prosperity in Metropolitan America." In *The New Suburban History*, ed. Kevin M. Kruse and Thomas J. Sugrue. Chicago: University of Chicago Press, 2006.

Frey, William H. "Census 2000." *Urban Land* (May 2002): 75–80.

———. "Melting Pot Suburbs: A Study of Suburban Diversity." In *Redefining Urban and Suburban America: Evidence from Census 2000*, vol. 1, ed. Bruce Katz and Robert E. Lang. Washington, D.C.: Brookings Institution Press, 2003.

Frey William H., and Alan Berube. "City Families and Suburban Singles: An Emerging Household Story." In *Redefining Urban and Suburban America: Evidence from Census 2000*, vol. 1, ed. Bruce Katz and Robert E. Lang. Washington, D.C.: Brookings Institution Press, 2003.

Frieden, Bernard. *The Environmental Protection Hustle.* Cambridge, Mass.: MIT Press, 1979.

Fulton, William. *The Reluctant Metropolis: The Politics of Urban Growth in Los Angeles.* Berkeley, Calif.: Solano Press Books, 1997.

Gaines, Donna. *Teenage Wasteland: Suburbia's Deadend Kids.* New York: Pantheon, 1991.

Galbraith, John Kenneth. *The Affluent Society.* Boston: Houghton Mifflin, 1976.

Galster George C., and E. Godfrey. "Racial Steering by Real Estate Agents in the U.S. in 2000." *Journal of the American Planning Association* 71 (2005): 251–265.

Galster, George, Royce Hanson, Michael Ratcliffe, Harold Wolman, Stephen Coleman, and Jason Freihage. "Wrestling Sprawl to the Ground—Defining and Measuring an Elusive Concept." In *Rates, Trends, Causes, and Consequences of Urban Land-Use Change in the United States*, ed. William Acevedo, Janis L. Taylor, et al. Professional Paper 1726. Reston, Va.: U.S. Geological Service, 2006.

Gans, Herbert. *The Levittowners: Ways of Life and Politics in a New Suburban Community.* New York: Vintage, 1967.

———. *The Urban Villagers.* New York: Free Press, 1962.

Garreau, Joel. *Edge City: Life on the New Frontier.* New York: Doubleday, 1991.

Geddes, Patrick. *Cities in Evolution.* London: Williams and Norgate, 1915.

Gertner, Jon. "Chasing Ground." *New York Times Magazine*, October 16, 2005.

Giddens, Anthony. *Central Problems in Social Theory.* London: Macmillan, 1979.

———. *The Consequences of Modernity.* Stanford, Calif.: Stanford University Press, 1990.

————. *The Constitution of Society: Outline of the Theory of Structuration.* Cambridge: Polity Press, 1984.

Glaeser Edward L., and Joseph Gyourko. *The Impact of Zoning on Housing Affordability.* Working Paper 8835. Cambridge, Mass.: National Bureau of Economic Research, 2002.

Goodman, Paul. *Communitas: Means of Livelihood and Ways of Life.* New York: Random House, 1960.

Gordon, Peter, and Harry Richardson. *The Case for Suburban Development.* Los Angeles: School of Urban and Regional Planning, UCLA, 1995.

Gordon, Tracy M. *Planned Developments in California: Private Communities and Public Life.* San Francisco: Public Policy Institute of California, 2004.

Goss, Jon. "The 'Magic of the Mall': An Analysis of Form, Function, and Meaning in the Contemporary Retail Built Environment." *Annals of the Association of American Geographers* 83 (1993): 18–47.

————. "Once-upon-a-Time in the Commodity World: An Unofficial Guide to Mall of America." *Annals of the Association of American Geographers* 89 (1999): 45–75.

Gottdiener, Mark. *Planned Sprawl: Public and Private Interests in Suburbia.* Beverly Hills, Calif.: Sage, 1977.

Gottmann, Jean. *Megalopolis: The Urbanized Northeastern Seaboard of the United States.* New York: Twentieth-Century Fund, 1961.

Graham, Simon, and S. Marvin. *Splintering Urbanism.* London: Routledge, 2001.

Gram-Hanssen, Kirsten, and Claus Bech-Danielsen. "House, Home and Identity from a Consumption Perspective." *Housing, Theory and Society* 21 (2004): 17–26.

Grant, Jill. *Planning the Good Community: New Urbanism in Theory and Practice.* New York: Routledge, 2006.

Habermas, Jürgen. *The Structural Transformation of the Public Sphere.* Cambridge, Mass.: MIT Press, 1989.

Hackworth, Jason. *The Neoliberal City: Governance, Ideology, and Development in American Urbanism.* Ithaca, N.Y.: Cornell University Press, 2007.

Hall, Peter. *Cities of Tomorrow.* Oxford: Blackwell, 1988.

————. "Global City-Regions in the 21st Century." In *Global City-Regions: Trends, Theory, Policy*, ed. A. J. Scott. New York: Oxford University Press, 2001.

Hall, Peter, and Kathy Pain. *The Polycentric Metropolis.* Sterling, Va.: Earthscan, 2006.

Harvey, David. *The Condition of Postmodernity.* Oxford: Blackwell, 1989.

————. "The New Urbanism and the Communitarian Trap: On Social Problems and the False Hope of Design." *Harvard Design Magazine* (Winter/Spring 1997): 68–69.

————. *Spaces of Global Capitalism: A Theory of Uneven Geographical Development.* New York: Verso, 2006.

————. *Spaces of Hope.* Berkeley: University of California Press, 2000.

————. *The Urbanization of Capital: Studies in the History and Theory of Capitalist Urbanization.* Baltimore: Johns Hopkins University Press, 1985.

Hayden, Dolores. *Building Suburbia: Green Fields and Urban Growth, 1820–2000.* New York: Pantheon, 2003.

————. *The Grand Domestic Revolution: A History of Feminist Designs for American Homes, Neighborhoods, and Cities.* Cambridge, Mass.: MIT Press, 1981.

Healy, Patsy. "An Institutional Model of the Development Process." *Journal of Property Research* 9 (1992): 33–44.

————. "Models of the Development Process: A Review." *Journal of Property Research* 8 (1992): 219–238.

Healy, Patsy, and Susan Barrett. "Structure and Agency in Land and Property Development Processes." *Urban Studies* 27 (1990): 89–104.

Heidegger, Martin. *Poetry, Language, Thought.* New York: Harper & Row, 1971.

Heller, Agnes. "Existentialism, Alienation, Postmodernism: Cultural Movements as Vehicles of Change in Patterns of Everyday Life." In *Postmodern Conditions*, ed. A. Milner, P. Thompson, and C. Worth. New York: Berg, 1990.

Hogen-Esch, Tom. "Urban Secession and the Politics of Growth: The Case of Los Angeles." *Urban Affairs Review* 36 (2001): 785.

Hornstein, Jeffrey M. *A Nation of Realtors: A Cultural History of the Twentieth-Century American Middle Class.* Durham, N.C.: Duke University Press, 2005.

Howe, Therese. "Ritz's Creighton Farms Opens for Home Sales." *Leesburg Today*, October 20, 2006.

Hugill, Peter J. "Good Roads and the Automobile in the United States, 1880–1929." *Geographical Review* 72 (1982): 327–349.

Jackson, Kenneth T. *Crabgrass Frontier: The Suburbanization of the United States.* New York: Oxford University Press, 1985.

Jackson, Peter. "Social Disorganization and Moral Order in the City." *Transactions, Institute of British Geographers* n.s. 9 (1984): 168–180.

Jacobs, Jane. *The Death and Life of Great American Cities.* New York: Random House, 1961.

Jameson, Fredric. *The Cultural Turn: Selected Writings on the Postmodern, 1983–1998.* London: Verso, 1998.

———. *Postmodernism; or, The Cultural Logic of Late Capitalism.* Durham, N.C.: Duke University Press, 1991.

Jonas, Andrew, and David Wilson, eds. *The Urban Growth Machine: Critical Perspectives Two Decades Later.* Albany: State University of New York Press, 1999.

Kaiser, Edward, and S. Weiss. "Public Policy and the Residential Development Process." *Journal of the American Institute of Planners* 36 (1970): 23–34.

Katz, Bruce. "Welcome to the 'Exit Ramp' Economy." *Boston Globe*, May 13, 2001, A19.

Katz, Peter, ed. *The New Urbanism: Toward an Architecture of Community.* New York: McGraw-Hill, 1994.

Kellner, Douglas. "Media Culture and the Triumph of the Spectacle." *Fast Capitalism* 1.1 (2005). http://www.uta.edu/huma/agger/fastcapitalism/1.1/kellner.html.

Kerwin, Kathleen. "A New Blueprint at Pulte Homes." *Business Week.* October 3, 2005, 76.

Kim, J. "Perceiving and Valuing Sense of Community in a New Urbanist Development: The Case of Kentlands." *Journal of Urban Design* 12 (2007): 203–230.

Kim, J., and R. Kaplan. "Physical and Psychological Factors in Sense of Community: New Urbanist Kentlands and Nearby Orchard Village." *Environment and Behavior* 36 (2004): 313–340.

Kirby, Andrew, Sharon L. Harlan, Larissa Larsen, Edward Hackett, Bob Bolin, Amy Nelson, Tom Rex, and Shapard Wolf. "Examining the Significance of Housing Enclaves in the Metropolitan United States of America." *Housing, Theory, and Society* 23 (2006):19–33.

Knorr-Cetina, Karin "Postsocial Relations: Theorizing Sociality in a Postsocial Environment." In *Handbook of Social Theory*, ed. G. Ritzer and B. Smart. London: Sage, 2001.

Knox, Paul L. "Capital, Material Culture, and Socio-Spatial Differentiation." In *The Restless Urban Landscape*, ed. P. L. Knox. Englewood Cliffs, N.J.: Prentice-Hall, 1993.

―――. "Schlock and Awe." *The American Interest* 2 (2007): 58–67.

―――. "Town Planning and the Internal Survival Mechanisms of Urbanized Capitalism." *Area* 13 (1981): 183–188.

―――. "Vulgaria: The Re-Enchantment of Suburbia." *Opolis* 1 (2005): 34–47.

Knox, Paul L., and Joe Cullen. "Planners as Urban Managers." *Environment & Planning A* 13 (1981): 885–898.

Knox, Paul L., and Kylie Johnson. "Learning from Las Vegas: Place Marketing and the Arts." *International Journal of the Arts in Society* 1 (2007): 47–54.

Knox, Paul L., and Linda McCarthy. *Urbanization.* 2nd ed. Upper Saddle River, N.J.: Prentice-Hall, 2005.

Knox Paul L., and Steven Pinch. *Urban Social Geography.* 5th ed. London: Pearson, 2006.

Kohn, Margaret. *Brave New Neighborhoods: The Privatization of Public Space.* New York: Routledge, 2004.

Kolson, K. *Big Plans: The Allure and Folly of Urban Design.* Baltimore: Johns Hopkins University Press, 2001.

Koolhaas, Rem, Stefano Boeri, Sanford Kwinter, Nadia Tazi, and Hans Obrist. *Mutations: Harvard Project on the City.* Barcelona: ACTAR; Bordeaux, France: Arc en rêve centre d'architecture, 2001.

Kotkin, Joel. "Suburbia: Homeland of the American Future." *Next American City* 11 (2006): 19–22.

Kowinski, William. *The Malling of America.* New York: Morrow, 1985.

Krieger, Alex. "The Costs—and Benefits? —of Sprawl." In *Sprawl and Suburbia*, ed. William Saunders. Minneapolis: University of Minnesota Press, 2005.

―――. "Whose Urbanism?" *Architecture* (November 1998): 72–75.

Kunstler, James H. *The Geography of Nowhere: The Rise and Decline of America's Man-Made Landscape.* New York: Simon & Schuster, 1993.

―――. *Home from Nowhere: Remaking Our Everyday World for the Twenty-first Century.* New York: Simon & Schuster, 1996.

Lake, Robert. "Spatial Fix 2: The Sequel." *Urban Geography* 16 (1995): 189–191.

Lang, Robert E. *Edgeless Cities: Exploring the Elusive Metropolis.* Washington, D.C.: Brookings Institution Press, 2003.

Lang, Robert E., and Edward Blakely. "In Search of the Real OC: Exploring the State of American Suburbs." *The Next American City* (August 2006): 16–18.

Lang, Robert E., and Karen A. Danielsen. "Monster Homes." *Planning* (May 2002): 2–7.

Lang, Robert E., and J. S. Hall. *The Sun Corridor: Planning Arizona's Megapolitan Area.* Tempe, Ariz.: Morrison Institute of Public Policy, 2006.

Lang, Robert E., and Paul L. Knox. "The New Metropolis: Rethinking Megalopolis." *Regional Studies* 42 (2008): in press.

Lang, Robert E., and Jennifer LeFurgy. *Boomburbs: The Rise of America's Accidental Cities.* Washington, D.C.: Brookings Institution Press, 2007.

Lang, Robert E., Jennifer LeFurgy, and Arthur C. Nelson. "The Six Suburban Eras of the United States." *Opolis* 2 (2006): 65–72.

Lang, Robert E., and Arthur C. Nelson. "America 2040: The Rise of the Megapolitans." *Planning* (January 2007): 5–12.

―――. *Beyond Metroplex: Examining Commuter Patterns at the Megapolitan Scale.* Cambridge, Mass.: Lincoln Institute for Land Policy, 2007.

Lang, Robert E., Thomas Sanchez, and Jennifer LeFurgy. *Beyond Edgeless Cities: A New Classification System for Suburban Business districts.* Washington, D.C.: The National Association of Realtors, 2006.

Lash, Scott, and John Urry. *Economies of Signs and Spaces: After Organized Capitalism.* Cambridge: Cambridge Policy Press, 1992.

Latham, Alan . "American Dreams, American Empires, American Cities." *Urban Geography* 25 (2004): 788–791.

Lauria, M. *Reconstructing Urban Regime Theory.* Beverly Hills, Calif.: Sage, 1996.

Leach, William. *Land of Desire, Merchants of Power, and the Rise of a New American Culture.* New York: Vintage, 1993.

Leitner, Helga, Jamie Peck, and Eric Sheppard, eds. *Contesting Neoliberalism: Urban Frontiers.* New York: Guilford Press, 2007.

Levine, Jonathan. *Zoned Out: Regulation, Markets, and Choices in Transportation and Metropolitan Land-Use.* Washington, D.C.: Resources for the Future, 2006.

Lewis, Peirce F. "The Galactic Metropolis." In *Beyond the Urban Fringe,* ed. R. H. Pratt and G. Macinko. Minneapolis: University of Minnesota Press, 1983.

Ley, David. "Co-Operative Housing as a Moral Landscape: Re-examining 'the Postmodern City.'" In *Place/Culture/Representation,* ed. J. Duncan and D. Ley. London: Routledge, 1993.

Logan, John, and Harvey Molotch. *Urban Fortunes: The Political Economy of Place.* Berkeley: University of California Press, 1987.

Loh, Sandra Tsing. "Kiddie Class Struggle." *Atlantic Monthly* 295 (June 2005): 114–116.

Low, Setha. *Behind the Gates: Life, Security, and the Pursuit of Happiness in Fortress America.* New York: Routledge, 2003.

Lucy, William, and David Phillips. *Tomorrow's Cities, Tomorrow's Suburbs.* Chicago: Planners Press, 2006.

Lund, H. "Testing the Claims of New Urbanism: Local Access, Pedestrian Travel, and Neighboring Behaviors." *Journal of the American Planning Association* 69 (2003): 414–429.

Lynch, Kevin. *Image of the City.* Cambridge, Mass.: MIT Press, 1960.

Lynd, R. S., and H. M. Lynd. *Middletown.* New York: Harcourt Brace Javanovich, 1956.

Maffesoli, Michael. *The Time of the Tribes: The Decline of Individualism in Mass Society.* London: Sage, 1996.

Mahar, Lisa. *American Signs: Form and Meaning on Route 66.* New York: Monacelli, 2002.

Mahler, Jonathan. "The Soul of the New Exurb," *New York Times Magazine,* March 27, 2005.

Mariani, Michele. "Consistent Performers." *Builder Magazine,* May 1, 2006.

Marin, Louis. *Utopics: The Semiological Play of Textual Spaces.* Amherst, N.Y.: Prometheus Books, 1990.

Marshall, Alex. *How Cities Work: Suburbs, Sprawl, and the Roads Not Taken.* Austin: University of Texas Press, 2003.

Marx, Leo. *The Machine in the Garden.* New York: Oxford University Press, 1964.

McCracken, G. *Culture and Consumption: New Approaches to the Symbolic Character of Consumer Goods and Activities.* Bloomington: Indiana University Press, 1988.

McGurr, Lisa. *Suburban Warriors: The Origins of the New American Right.* Princeton, N.J.: Princeton University Press, 2001.

McKenzie, Evan. "Constructing The *Pomerium* in Las Vegas: A Case Study of Emerging Trends in American Gated Communities." In *Gated Communities,* ed. Rowland Atkinson and Sarah Blandy. New York: Routledge, 2006.

———. *Privatopia: Homeowner Associations and the Rise of Residential Private Government.* New Haven: Yale University Press, 1994.

McPherson, Miller, Lyn Smith-Lovin, and Matthew Brashears. "Social Isolation in America: Changes in Core Discussion Networks over Two Decades." *American Sociological Review* 71 (2006): 353–375.

Mencken, H. L. *Prejudices: Sixth Series.* New York: Knopf, 1927.

Mitchell, Don. *Cultural Geography: A Critical Introduction.* Oxford: Blackwell, 2000.

———. "Landscape and Surplus Value: The Making of the Ordinary in Brentwood, California." *Environment & Planning D: Society and Space* 12 (1994): 7–30.

Mitchell, W.J.T., ed. *Landscapes and Power.* Chicago: University of Chicago Press, 1994.

Mohney David, and Keller Easterling, eds. *Seaside: Making a Town in America.* Princeton, N.J.: Princeton Architectural Press, 1991.

Molotch, Harvey. "The City as a Growth Machine: Toward a Political Economy of Place." *American Journal of Sociology* 82 (1976): 309–310.

Monahan, Torin. "Electronic Fortification in Phoenix: Surveillance Technologies and Social Regulation in Residential Communities." *Urban Affairs Review* 42 (2006): 169–192.

Morin, Richard, and Steven Ginsberg. "Painful Commutes Don't Stop Drivers." *Washington Post*, February 13, 2005, A01.

Morris, Douglas, *It's a Sprawl World After All: The Human Cost of Unplanned Growth.* Gabriola Island, B. C., Canada: New Society Publishers, 2005.

Mumford, Lewis. *The City in History: Its Origins, Its Transformations, and Its Prospects.* New York: Harcourt, Brace & World, 1961.

———. *The Culture of Cities.* London: Secker and Warburg, 1938.

———. "Regions—To Live In." *Survey* 54 (1925): 150–152.

———. *The Story of Utopias.* London: Harrap, 1923.

Myers, Dowell, and Elizabeth Gearin. "Current Preferences and Future Demand for Denser Residential Environments." *Housing Policy Debate* 12 (2001): 633–659.

Nasar, J. "Does Neotraditional Development Build Community?" *Journal of Planning Education and Research* 23 (2003): 58–68.

National Association of Homebuilders Economics Group. *What 20th Century Home Buyers Want.* Washington, D.C.: National Association of Home Builders, 2001.

Nelson, Arthur C. "Characterizing Exurbia." *Journal of Planning Literature* (1992): 350–368.

———. "How Do We Know Smart Growth When We See It?" In *Smart Growth: Form and Consequences*, ed. Terry Szold and Armando Carbonell. Cambridge, Mass.: Lincoln Institute of Land Policy, 2002.

Nelson, Arthur C., and Robert E. Lang. "The Next 100 Million." *Planning* (2007): 2–4.

Nelson, Robert H. *Private Neighborhoods and the Transformation of Local Government.* Washington, D.C.: Urban Institute Press, 2005.

Newman, Peter, and Jeffrey Kenworthy. *Cities and Automobile Dependence.* Aldershot, Eng.: Gower Publishing, 1989.

Newton, Ken. "Conflict Avoidance and Conflict Suppression." In *Urbanization and Conflict in Market Societies*, ed. Kevin Cox. London: Methuen, 1978.

Nolen, John. *New Towns for Old: Achievements in Civic Improvement in Some American Small Towns and Neighborhoods.* 1927; reprint, Amherst: University of Massachusetts Press, 2005.

Ogborn, M., and C. Philo. "Soldiers, Sailors and Moral Locations in Nineteenth-Century Portsmouth." *Area* 26 (1994): 221–231.

Orfield, Myron. *American Metropolitics.* Washington, D.C.: Brookings Institution Press, 2002.

Ostrom, E. "Metropolitan Reform: Propositions Derived from Two Traditions." *Social Science Quarterly* 53 (1972): 474–493.

Pahl, Ray. "Urban Social Theory and Research." *Environment and Planning A* 1 (1969): 143–153.

Paterson, Mark. *Consumption and Everyday Life.* London: Routledge, 2006.

Peck, Jamie, and Adam Tickell. "Neoliberalizing Space." In *Spaces of Neoliberalism: Urban Restructuring in North America and Western Europe*, ed. Neil Brenner and Nik Theodore. Oxford: Blackwell, 2002.

Pierce, Emmet. "Associations Are Too Quick on the Trigger, Critics Claim." *San Diego Union-Tribune,* April 10, 2005.

Pinder, David. *Visions of the City.* New York: Routledge, 2005.

Ploszajska, T. "Moral Landscapes and Manipulated Spaces: Gender, Class and Space in Victorian Reformatory Schools." *Journal of Historical Geography* 20 (1994): 413–429.

Popper, Deborah E., Robert E. Lang, and Frank J. Popper. "From Maps to Myth: The Census, Turner, and the Idea of the Frontier." *Journal of American and Comparative Popular Cultures* 23 (2001): 91–102.

Postrel, Virginia. *The Substance of Style.* New York: HarperCollins, 2003.

Putnam, Robert. *Bowling Alone: The Collapse and Revival of American Community.* New York: Simon & Schuster, 2000.

Real Estate Research Corporation. *The Costs of Sprawl: Environmental and Economic Costs of Alternative Residential Development Patterns at the Urban Fringe.* Vol. 1, *Detailed Cost Analysis*; Vol. 2, *Literature Review*; Vol. 3, *Executive Summary.* Washington, D.C.: U.S. Government Printing Office, 1974.

Reich, Robert. "Secession of the Successful." *New York Times Magazine,* January 20, 1991, 16–18.

———. *The Work of Nations: Preparing Ourselves for 21st-Century Capitalism.* New York: Knopf, 1991.

Relph, Edward. *The Modern Urban Landscape.* London: Croom Helm, 1987.

———. *Place and Placelessness.* London: Pion, 1976.

Riesman, David. *The Lonely Crowd.* New Haven: Yale University Press, 1950.

Ritzer, George. *Enchanting a Disenchanted World.* 2nd edition. Thousand Oaks, Calif.: Pine Forge Press, 2005.

Robbins, Edward. "New Urbanism." In *Shaping the City: Studies in History, Theory, and Urban Design*, ed. Edward Robbins and Rodolphe El-Khoury. New York: Routledge, 2004.

Rohe, W. M., and H. L. Watson, eds. *Chasing the American Dream. New Perspectives on Affordable Homeownership.* Ithaca, N.Y.: Cornell University Press, 2007.

Romig, Kevin. "Upper Sonoran Lifestyles: Gated Communities in Scottsdale, Arizona." *City and Community* 4 (2005): 67–86.

Ross, Andrew. *The Celebration Chronicles: Life, Liberty, and the Pursuit of Property Value in Disney's New Town.* New York: Ballantine Books, 1999.

Rothman, Hal. *Neon Metropolis: How Las Vegas Started the Twenty-first Century.* London: Routledge, 2002.

Rowe, Colin. "Collage City." *Architectural Review* (August 1975): 65–91.

Rudofsky, Bernard. *Architecture Without Architects: An Introduction to Non-pedigreed Architecture.* New York: Museum of Modern Art, 1964.

Rusk, David. *Cities Without Suburbs.* Washington, D.C.: Woodrow Wilson Center Press, 1995.

Rutheiser, C. *Imagineering Atlanta.* New York: Verso, 1996.

Rybcznski, Witold. *City Life: Urban Expectations in a New World.* New York: Scribner, 1995.

Sack, Robert. "The Consumer's World: Place as Context." *Annals, Association of American Geographers* 78 (1988): 642–665.

Sanchez, Thomas W., Robert E. Lang, and Dawn M. Dhavale. "Security versus Status? A First Look at the Census's Gated Community Data." *Journal of Planning Education and Research* 24 (2005): 281–291.

Saunders, William S., ed. *Sprawl and Suburbia.* Minneapolis: University of Minnesota Press, 2005.

Schmitz, Adrienne, R. Bird Anderson, Leslie Holst, Wayne Hyatt, Frederick Jarvis, Jennifer LeFurgy, Ehud Mouchly, Sam Newberg, Peter Smirniotopoulos, and Meghan Welsch. *Residential Development Handbook.* Washington, D.C.: Urban Land Institute, 2004.

Schmitz, Adrienne, Pam Engebretson, Frederick Merrill, Sarah Peck, Robert Santos, Kristin Shewfelt, Debra Stein, John Torti, and Marilee Utter. *The New Shape of Suburbia: Trends in Residential Development.* Washington, D.C.: Urban Land Institute, 2003.

Schor, Juliet. *The Overspent American: Upscaling, Downshifting, and the New Consumer.* New York: Basic Books, 1998.

Schuyler, David. *The New American Landscape: The Redefinition of City Form in Nineteenth-Century America.* Baltimore: Johns Hopkins University Press, 1986.

Schwarzer, Mitchell. "The Spectacle of Ordinary Building." *Harvard Design Magazine* 12 (2000): 13–19.

Sennett, Richard. *The Conscience of the Eye: The Design and Social Life of Cities.* New York: Knopf, 1990.

———. "The Search for a Place in the World." In *Architecture of Fear*, ed. Nan Ellin. New York: Princeton Architectural Press, 1997.

———. *The Uses of Disorder. Personal Identity and City Life.* New York: Random House, 1970.

Severgnini, Beppe. *Ciao America: An Italian Discovers the U.S.* New York: Broadway Books, 2001.

Shames, L. "What a Long, Strange (Shopping) Trip It's Been: Looking Back at the 1980s." *Utne Reader* (September/October 1989): 117–121.

Shields, Rob. "The Individual, Consumption Cultures, and the Fate of Community." In *Lifestyle Shopping: The Subject of Consumption*, ed. R. Shields. New York: Routledge, 1992.

Siegan, Bernard H. *Land Use without Zoning.* Lexington, Mass.: Lexington Books, 1972.

Sies, Mary. *The Suburban Ideal: A Cultural Strategy for Modern American Living.* Philadelphia: Temple University Press, 2000.

Simmons, Patrick, and Robert E. Lang. "The Urban Turnaround." In *Redefining Urban and Suburban America*, ed. Bruce Katz and Rob Lang. Washington, D.C.: Brookings Institution Press, 2003.

Smith, Henry Nash. *Virgin Land: The American West as Symbol and Myth.* Cambridge, Mass.: Harvard University Press, 1971.

Smith, Neil. *The New Urban Frontier.* New York: Routledge, 1996.

Soja, Edward. *Postmetropolis: Critical Studies of Cities and Regions.* Oxford: Blackwell, 2000.

———. *Postmodern Geographies.* London: Verso, 1989.

———. "The Socio-Spatial Dialectic." *Annals of the Association of American Geographers* 70 (1980): 207–225.

Sorkin, Michael. "Brand Aid; or, The Lexus and the Guggenheim (Further Tales of the Notorious B.I.G.ness)." *Harvard Design Magazine* 17 (2002): 1–5.

———, ed. *Variations on a Theme Park.* New York: Hill and Wang, 1992.

Spectorsky, Auguste. *The Exurbanites.* Philadelphia: Lippincott, 1955.

Squires, Gregory D. "Urban Sprawl and the Uneven Development of Metropolitan America." In *Urban Sprawl: Causes, Consequences, and Policy Responses,* ed. Gregory D. Squires. Washington, D.C.: Urban Institute Press, 2002.

Stein, M. *The Eclipse of Community.* New York: Harper & Row, 1960.

Stilgoe, John. *Borderlands: Origins of the American Suburb, 1820–1939.* New Haven: Yale University Press, 1990.

Strasser, Susan. *The Making of the American Mass Market.* New York: Pantheon, 1989.

Sudjic, Deyan. *The Edifice Complex.* New York: Penguin, 2005.

———. *The 100-Mile City.* London: Andre Deutsch, 1991.

Sutton, Paul C. "A Scale-Adjusted Measure of Urban Sprawl Using Nighttime Satellite Imagery." *Remote Sensing of Environment* 86 (2003): 353–369.

Swanstrom, T., C. Casey, R. Flack, and P. Dreier. *Pulling Apart: Economic Segregation among Suburbs and Central Cities in Major Metropolitan Areas.* Washington, D.C.: Brookings Institution Metropolitan Policy Program, 2004.

Swyngedouw, Eric. "Exit 'Post'—The Making of 'Glocal' Urban Modernities." In *Future City,* ed. S. Read, J. Rosemann, and J. van Eldijk. London: Spon Press, 2005.

Sykes, Simon. "Profile: Living Las Vegas." *Urban Land* 64 (2005): 23–24.

Tafuri, Manfredo, and Francesco Co. *Modern Architecture.* New York: Rizzoli, 1986.

Talen, Emily. *New Urbanism and American Planning: The Conflict of Cultures.* New York: Routledge, 2005.

———. "Sense of Community and Neighborhood Form: An Assessment of the Social Doctrine of New Urbanism." *Urban Studies* 36 (1999): 1361–1379.

Tarlock, Dan. "Toward a Revised Theory of Zoning." In *Land Use Controls Annual,* ed. F. S. Bangs. Chicago: American Society of Planning Officials, 1972.

Teaford, Jon. *The Metropolitan Revolution.* New York: Columbia University Press, 2006.

Theobald, David M. "Landscape Patterns of Exurban Growth in the USA from 1980 to 2020." *Ecology and Society* 10 (2005).

———. "Land-Use Dynamics beyond the American Urban Fringe." *Geographical Review* 91 (2001): 544–564.

Thoreau, Henry D. *Walden.* 1854. Reprint, New York: Holt, Rinehart, and Winston, 1961.

Thrift, Nigel. "Transurbanism." *Urban Geography* 25 (2004): 724–734.

Tiebout, C. "A Pure Theory of Local Public Expenditure." *Journal of Political Economy* 64 (1956): 416–424.

Tönnies, Ferdinand. *Community and Society.* Trans. C. P. Loomis. 1887. Reprint, New York: Harper, 1963.

Tschumi, Bernard. *Architecture and Disjunction.* Cambridge, Mass.: MIT Press, 1994.

Tuan, Yi-Fi. *Passing Strange and Wonderful: Aesthetics, Nature, and Culture.* Washington, D.C.: Island Press, 1993.

Turner, Frederick J. *The Frontier in American History.* New York: Henry Holt, 1920.

Twitchell, James. *Branded Nation: The Marketing of Megachurch, College, Inc., and Museumworld.* New York: Simon & Schuster, 2004.

———. *Living It Up. America's Love Affair with Luxury.* New York: Simon & Schuster, 2002.

United Nations Centre for Human Settlements. *Global Report on Human Settlements, 2001.* London: Earthscan, 2001.

Upton, D. "The Tradition of Change." *Traditional Dwellings and Settlements Review* 5 (1993): 9–16.

Urban Land Institute. *Great Planned Communities.* Washington, D.C.: Urban Land Institute, 2002.

———. *Housing for Emerging Niche Markets.* InfoPacket No. 3007. Washington, D.C.: Urban Land Institute, 2006.

U.S. Department of Housing and Urban Development. *The State of the Cities.* Washington, D.C.: U.S. Government Printing Office, 1999.

U.S. Department of Housing and Urban Development Office of Policy Development and Research. *American Housing Survey for the United States 2005.* Washington, D.C.: U.S. Department of Housing and U.S. Department of Commerce, 2006.

Vance, James E. Jr. *This Scene of Man: The Role and Structure of the City in the Geography of Western Civilization.* New York: Harper's College Press, 1977.

Veblen, Thorstein. *The Theory of the Leisure Class.* New York: Macmillan, 1899.

Venturi, Robert. *Complexity and Contradiction in Architecture.* New York: Museum of Modern Art, 1966.

Venturi, Robert, Denise Scott Brown, and Steven Izenour. *Learning from Las Vegas: The Forgotten Symbolism of Architectural Form.* Cambridge, Mass.: MIT Press, 1977.

Vicino, T. J., B. Hanlon, and J. Short. "Megalopolis 50 Years On: The Transformation of a City Region." *International Journal of Urban and Regional Research* 31 (2007): 344–367.

Warner, Sam Bass Jr. *Streetcar Suburbs.* Cambridge, Mass.: Harvard University Press, 1962.

Warner, W. L., and P. S. Lunt. *The Social Life of a Modern Community.* New Haven: Yale University Press, 1941.

Weibel, Peter. "Las Vegas, The City—A Place of Consumption in the Post-industrial Information Society." In *The Magic Hour: The Convergence of Art and Las Vegas*, ed. Alex Farquharson. Ostfildern, Ger.: Hatje Cantz, 2001.

Weiss, Jonathan, and Randy Lowell. "Supersizing Religion: Megachurches, Sprawl, and Smart Growth." *St. Louis University Public Law Review* 21 (2002): 313–329.

Weiss, Marc A. *The Rise of the Community Builders: The American Real Estate Industry and Urban Land Planning.* New York: Columbia University Press, 1987.

White, Richard, and Patricia N. Limerick. *The Frontier in American Culture.* Berkeley: University of California Press, 1994.

Whyte, William. *The Organization Man.* New York: Doubleday, 1956.

Wilbur, Victoria R. *ULI Forum Report.* Washington, D.C.: Urban Land Institute, 2002.

Williams, Raymond. *The City and the Country.* London: Chatto and Windus, 1973.

Wilson, Elizabeth. "The Rhetoric of Urban Space." *New Left Review* 209 (1995): 146–160.

Wilson, William H. *The City Beautiful Movement.* Baltimore: Johns Hopkins University Press, 1989.

Wilson-Dogenges, Georjeanna. "An Exploration of Sense of Community and Fear of Crime in Gated Communities." *Environment and Behavior* 32 (2000): 597–611.

Woodcock, G. *Anarchism: A History of Liberation Ideas and Movements.* New York: Meridian Books, 1962.

Woods, J. " 'Build, Therefore, Your Own World': The New England Village as Settlement Ideal." *Annals of the Association of American Geographers* 81 (1991): 32–50.

Zukin, Sharon. *Landscapes of Power: From Detroit to Disney World.* Berkeley: University of California Press, 1991.

INDEX

Abu-Lughod, Janet, 103
"Achievers," 137–139, *137, 139, 141,*
   145, 158
acquisitions, 67, 68–69, 72
"active adult" communities, 56, 72, 83,
   84–86, 122
actors, 6–9, 11–13, 66, 90. *See also types*
   of actors, e.g., intellectuals, design
   professionals
Adams, James Truslow, 22
Adelson, Sheldon, 157
advertising, 5, 11, 44, 60, 76; and degenerate utopias, 98–99, 101; and material
   culture, 134, 136, 143–144, 149,
   151–152
aesthetics, 5–6; and degenerate utopias,
   89, 91–92, 95, 97; and developers'
   utopias, 79–83; in historical context of
   suburbia, 17, 24, 33–34; and material
   culture, 142; and privatism politics,
   117, 126–127; and Vulgaria,
   159–160
affect, 34, 149, 155, 159–162
affluence, 4, 10–11; and degenerate
   utopias, 98, 107–108; and developers'
   utopias, 74–86; in historical context of
   suburbia, 14, 19–20, 29; and material
   culture, 133, 135–136, 140, 143,
   146–147, 149; and New Metropolis,
   45, 56–65, *60, 61, 62, 63, 64, 65*; and
   privatism politics, 113, 121, 125, 127;
   and Vulgaria, 158, 162, 170
Affluentials, 147–148
"affluenza," 136, 161, 173
affordable housing, 45, 84, 126–127, 159

African Americans, 56, 75, 94
age diversity, 8, 75, 105, 122
agents. *See* actors
Ahwahnee Principles, 100, 104–105
Alexander, Christopher, 96, 106
Alexandria (Va.), *51,* 104, 117
Aliante (Nev.), 85
Ambler, Euclid v., 26–28
amenities/services, 9; and degenerate
   utopias, 90, 92–93, 98; and
   developers' utopias, 73, 83–84, 86–88;
   in historical context of suburbia,
   16–17, 20, 22, 31; and material
   culture, 138, 143, 145,
   149–150, 152; and New Metropolis,
   52, 59; and privatism politics,
   113–116, 118, 125, 128; and
   Vulgaria, 157, 169–170
American Builders, Inc., 70
American Community Survey, 35
American Dream, 4, 7, 19–23, 25, 29, 56,
   76, 80, 133–135, 142, 159, 161, 164
American Dream Extreme, 59–60, 76–79,
   161–162, 173
American Farmland Trust, 32, 129
American Housing Survey, 80
American Institute of Architects,
   191n60
American Planning Association, 106, 129,
   153, 188n13
American Renaissance, 11, 14–15, 18,
   32, 92
Americans for Limited Government, 124
Anaheim (Calif.), 47, 56
Anthem (Ariz.), 84

ABOUT THE AUTHOR

Paul Knox is University Distinguished Professor and Senior Fellow for International Advancement at Virginia Tech, where he was dean of the College of Architecture and Urban Studies from 1997 to 2006. He has published widely on urban social geography and economic geography. Recent books include *Urbanization* (with Linda McCarthy), *Urban Social Geography* (with Steven Pinch), *Places and Regions in Global Context* (with Sallie Marston and Diana Liverman), *The Geography of the World-Economy* (with John Agnew and Linda McCarthy), and *Design Professionals and the Built Environment* (edited, with Peter Ozolins).